AF866159

CHRIST EVANGELICAL BIBLE INSTITUTE

INTELLIGENT EVOLUTION

VOLUMES ONE & TWO

Rev. Joseph Adam Pearson, Ph.D.

Copyright © 2022 by Rev. Joseph Adam Pearson, Ph.D.

All rights reserved.

This work is a revision of the following earlier versions
registered with the United States Copyright Office:

Copyright © 2021 (TX0009070948) by Joseph Adam Pearson
Copyright © 2020 (TX0008869477) by Joseph Adam Pearson
Copyright © 2018 (TX0008566448) by Joseph Adam Pearson
Copyright © 2017 (TX0008361069) by Joseph Adam Pearson
Copyright © 2016 (TX0008239803) by Joseph Adam Pearson

Hardcover Identifier:

ISBN-13: 9781734294774

Published by
Christ Evangelical Bible Institute
(SAN: 920-3753)
Dayton, Tennessee

Last edited on May 21, 2022

About *Intelligent Evolution*

The apprehension of all authors is to be misunderstood and, as a result of being misunderstood, to have their works misjudged. This book on *intelligent evolution* will only be misjudged when it is misunderstood, and it will only be misunderstood when readers choose to skim through it and not to think through and learn from it.

A theoretical expedition like *Intelligent Evolution* can only be fruitful if its readers are grounded in the knowledge of Christ Jesus as Savior and the *only-begotten* Son of God at the same time that their minds are open to additional invisible realities and unseen possibilities.

Intelligent Evolution is a study of abstract concepts that relate to cosmic evolution, biological evolution, and consciousness evolution under the governance of the Creator-God.

As an explorer of inner space through Christ Jesus since early childhood, the present author has not only learned to ignore the demonic illusions that exist in inner space but also to give no credence to the false glories of spiritual darkness that seem real there as well.

Igniting God's interest in us does not depend on how educated or uneducated we are. Concerning how educated or uneducated we are, God's grace and mercy are extended to each one of us because of His own nature and not because of our own individual or collective natures and experiences (except for our forgiving natures and salvation experiences). Each human being should recognize that, relative to the Creator-God, he or she is pitiful and pitiable as well as a wise fool; however, that recognition alone is insufficient to spark God's interest in us. What sparks God's interest in us is: (1) that we are His errant creation; and (2) the degree to which we seek His approval. *For example,* trying to ram our doctrinal beliefs down the throats of others does not spark God's interest in us; to the contrary, God prefers to let us stew in our own juices when we seek to control others. In order to seek God's approval, authentic Christians should be willing to learn not only what Christ Jesus loves in order to demonstrate it in thought, feeling, desire, deed, word, and attitude but also what Christ Jesus hates in order to avoid indulging or practicing it.

It is the author's hope that this book on *intelligent evolution* will not only ignite the interest and approval of the Creator-God but also the interest and approval of His people.

VOLUME ONE
Table of Contents

A Note to the Readers

As used in this book, *KJV* is an abbreviation for the public domain *King James Version* of the Holy Bible. To ensure their accuracy throughout this book, all paraphrases of the public domain *King James Version* of the Holy Bible were finalized only after first checking: (1) the Masoretic Hebrew text of the Tanakh (the Jewish Bible) for accuracy of passages from the *KJV Old Testament;* and (2) the earliest Greek text extant for accuracy of passages from the *KJV New Testament.* Additionally, to enhance readability of the public domain *KJV* text, the present author has changed words like *hath*, *thou*, and *ye* to their modern equivalents.

Although God the Father (i.e., the *Lord God Almighty*) and God the Son (i.e., the *Lord Jesus Christ*) are consubstantially united in the Godhead along with God the Holy Spirit, in order to distinguish *God the Father* from *God the Son*, an upper case "H" is used for personal pronouns specifically referring to *God the Father (He*, *His*, and *Him)* and a lower case "h" is used for personal pronouns specifically referring to *God the Son (he, his, and him).*

Most transliterated Hebrew and Greek words referenced within the text of this book (Volumes One and Two) are noted by their respective numbers [in brackets with a preceding "H" for Hebrew or "G" for Greek] from the *Dictionary of the Hebrew Bible* and the *Dictionary of the Greek Bible* found in *Strong's Exhaustive Concordance of the Bible* by James Strong (Copyright 1890), Crusade Bible Publishers, Inc., Nashville.

Whenever the title *God* is used in this book, the reader should assume that it is referring solely to: (1) the God of the Holy Bible — who is the *Lord God Almighty* or *Yahweh* (YHWH); (2) the one true and only real Creator-God, Creator-Evolver, and Creator-Savior; (3) *His* tripartite nature; (4) *His* sevenfold Spirit; and (5) *His* various facets.

Although the Creator-God does not possess a human gender, there are no apologies for the use of the male pronouns *He, His, and Him* when referring to the Lord God Almighty in this book for the following reason: In general, certain words in theology and philosophy are capitalized to show that they represent qualities and characteristics that transcend human understanding and experience. This includes the pronouns *He*, *His*, and *Him* and even the word *God* itself. *She* and *Her* are not used in this book when referring to the Creator-God because many people, if not most, have a tendency to confuse the use of female pronouns with advocating Wicca and other pagan cults that worship the Mother-Goddess — such as those devoted to Aphrodite (Venus), Artemis (Diana), Astarte, Cybele, Hecate, Ma'at, Morrigan, etc.

For the sake of clarity, when the author of *Intelligent Evolution* uses the phrase *the present author* in this book, he is referring to himself and not to some other author or source.

About the Author

It seems to me that I have lived my entire life believing that thoughts are things and that things are thoughts. *For example,* I remember a recurring dilemma throughout most of my childhood concerning the meaning of "exit" and "entrance" signs. Often, I had to pause at a door with such signage and think, "Am I exiting the store in order to enter the world or am I exiting the world in order to enter the store?" I often needed to look at the direction in which the doors would swing in order to solve the problem. This dilemma occurred regularly. As I saw it, life was *only* filled with conceptual puzzles that needed to be figured out. Now, as a senior adult, door signage continues to pose similar questions that I must ask myself (and answer correctly) before I act.

As a child, I often laughed when I fell. I thought it funny that the cumbersome body in which I found myself could be so clumsy and unaware of its surroundings or that its nervous system could be so incapable of making right decisions relative to the direction of its movements. I still laugh for similar reasons. Although I could write at length about many related things at this juncture, it is sufficient for me to state that, because I found the world to be an inhospitable place at an early age, it was easy for me to learn to dissociate myself from it. I have always felt, and still feel, like a stranger in a strange land. I have always felt, and still feel, that physicality, or corporeality, is alien to me and that I am an alien in it. As a result, throughout my entire life, I have always made a distinction between *physical existence (corporeal existence)* and *spiritual being (incorporeal being).*

Throughout my life, words, phrases, and statements have come to me from out of nowhere. *For example,* I remember walking home one day in 1966 and inwardly hearing: "Time is a sequence of related events." Every word and image that I received over the years, I would ponder and reflect on, often for decades. As I matured, I came to understand and accept that I had a susceptibility, or sensitivity, to external words and images from otherworldly sources.

I am very grateful for my mentoring as a young person by an aunt who had a substantial understanding of Christian metaphysics. She posed just the right questions to me about who I thought I am and who I really *am*. As a preteen, I remember her telling me to look at myself in her wall mirror. She asked me if the image in the mirror represented who I really *am*. I remember her telling me that it did not and why it did not. We met regularly to explore together who and what I was, and am, in God through Christ Jesus. During her tutelage, I became very comfortable with the concepts and language of Christian metaphysics, comparing and contrasting such concepts as *corporeality versus spirituality*, *absolute truth versus relative truth,* and *statements of existence versus declarations of being*.

As a young person, I loved traditional children's Sunday School. And I was a Vacation Bible School (VBS) junkie: During the summers, I would attend the Baptist VBS, Lutheran VBS, Methodist VBS, and Presbyterian VBS for two weeks each to study the Bible, memorize Bible verses, and work on Bible-related crafts. I also attended Bible Camp in Mukwonago, Wisconsin during the summers. I loved — and still love — reading, studying, and comprehending the Holy Bible and using the spiritual truths that it contains as a filter through which to view the world, its reality, and its unreality.

I remember deciding as a sophomore in high school what I wanted to do with my life: I wanted to become a biology teacher, a pastor, and an author.

In order to help fulfill my goals, I majored in biology at Loyola University in Chicago. My favorite science courses included: comparative embryology of vertebrates, comparative anatomy of vertebrates, physiology, histology, genetics, physics, and organic chemistry. In addition to science courses, I took various elective courses in world religions, Aristotelian logic and ethics, and metaphysics. I distinctly remember that my metaphysics professor, an ex-Jesuit, hated my written compositions because I always tried to link metaphysics to Christianity. I now understand that, although he believed in the existence of an invisible reality, he thought of it only as an intellectual reality and not a spiritual one. Like so many people today, he did not recognize that metaphysics is not only a legitimate branch of academic philosophy but also a legitimate branch of academic theology.

I have always enjoyed reading books directly and indirectly related to metaphysics — like Immanuel Kant's *Prolegomena to any Future Metaphysics* (1783) and Walter Haushalter's *Mrs. Eddy Purloins from Hegel* (1936). Today, I still read such works. *For example,* I have recently finished reading Friedrich Nietzsche's *Also sprach Zarathustra* (1885). I read the German original side-by-side with an English translation *(Thus Spoke Zarathustra)* to see if they really were the same book. (Because the two languages do not possess the same nuances of word meaning, I concluded that they are not *exactly* the same.)

After earning my Bachelor of Science degree in biology at Loyola University (Chicago) in 1969, I remained at Loyola for an additional two years to earn a Master of Science in biology with an emphasis in cell biology. Serving as a graduate teaching assistant in the Department of Biology at Loyola permitted me to finance my graduate studies: I especially enjoyed teaching human histology laboratory sections while I was there. During my Junior and Senior years as an undergraduate as well as during my graduate years at Loyola, I also worked as an electron microscopist in the Department of Oral Histology at the University of Illinois Dental School.

After receiving my Master of Science degree in 1971, I became a high school biology teacher at a prestigious, all-boys college preparatory school where I taught for two years. I then served for two years on the faculty as a Research Associate in the Department of Ophthalmology at the University of Illinois Medical Center, where I first-authored and co-authored many scientific papers in reputable, refereed (i.e., peer-reviewed) journals under my birth name of

Joseph Vlchek (J.K. Vlchek). While working as a Research Associate, I entered a doctoral program as a graduate student in the Department of Anatomy at the University of Illinois Medical School (the Abraham Lincoln School of Medicine). While in that program, I took advanced human anatomy, advanced human physiology, and advanced human histology. During that time, I also began to teach "Structure and Function of the Human Body," "Evolution, Genetics, and Development," and "Scientific Inquiry" as an adjunct faculty member in the Department of Natural Science at the Lewis Towers Campus of Loyola University.

Although I continued adjunct teaching at Loyola for many years, I left the University of Illinois Medical Center to take a full time teaching position with the City Colleges of Chicago in the Department of Biology at Kennedy-King College, where I taught human anatomy and physiology full time for eight years to students of medical education (primarily nursing students). Because the Department of Anatomy at the University of Illinois permitted only full time status for its doctoral students at the time, I matriculated into a doctoral program at the University of Chicago in its Department of Biology with the endorsement of the distinguished cell biologist, Dr. Hewson Swift, in whose laboratory I had conducted my research for my Master's thesis while at Loyola. At the University of Chicago, I took courses in biochemistry, lipoproteins and enzyme kinetics, and cell biology. Incidentally, the biochemistry course at the University of Chicago was the most difficult course I have ever taken. We covered the 1,000-page eighth edition of *Principles of Biochemistry* by Albert L. Lehninger in nine weeks, and students were responsible for all formulas, equations, and molecular structures in the book.

Eventually, I decided that I knew all that I needed to know for future independent learning in the content area of biology. I became more intrigued and challenged by the presentation of information to enhance its assimilation and accommodation by learners. So, in 1981, I left everything in Illinois to move to Arizona: (1) to enter a doctoral program in education at Arizona State University with an emphasis in teacher education, language, literacy, linguistics, and statistical analysis as well as (2) to teach for the Maricopa County Community College District, where I served full time as: (a) biology and chemistry faculty at South Mountain Community College for five years; (b) lead professor in human anatomy and physiology (as well as Biology Department Chairperson) at Scottsdale Community College for ten years; and (c) founding instructional dean at the Red Mountain Campus of Mesa Community College (MCC) and director of MCC's Extended Campus for a total of ten years. Altogether, I worked in the Maricopa County Community College District for twenty-five years. During that time, I earned my Doctor of Philosophy (Ph.D.) from Arizona State University in 1988 with a dissertation entitled *Testing the Ecological Validity of Student-Generated versus Teacher-Provided Postquestions in Reading College Science Text* (1988). I am pleased that my research findings were accepted for publication in the highly respected, refereed *Journal of Research in Science Teaching* in 1991.

Throughout my life, I have always multi-tasked and led double professional lives. *For example,* during the last ten years of the time that I worked for the Maricopa Community College District, I also served as Senior Pastor for Healing Waters Ministries in Tempe, Arizona. Additionally, for the past twenty-five years (1996-2021), I have served as International President and Chief Executive Officer of Christ Evangelical Bible Institute (CEBI), which has thriving branch campuses in India, Pakistan, and Tanzania. In that capacity, I have been responsible for developing, designing, and deploying Bible curriculum as well as for in-servicing the various branch campus administrators, ministerial students, and local pastors. At the time of this writing (2021), I am still serving as International President and CEO of CEBI as well as teaching online Bible courses.

I believe strongly that after we are saved, and at the same time we are being sanctified, our individual actions on Earth are part of an "application" for the jobs that we will each hold during Christ Jesus' millennial reign on Earth. My greatest goal is to be one of the many committed Christian educators who will be teaching throughout *the Millennium*. It is my hope that I will be able to use this book as a textbook for students of Christian metaphysics during that period of time.

When I was three years of age, I remember someone from Heaven telling me during an afternoon nap what my specific purpose for being on Earth is. I was also told that when I awoke I would not remember the specific purpose but that I would remember that I had been told. When I awoke from my nap that day, it was exactly so: I did not remember my specific purpose, but I did remember that I had been told. I suspect that the writing of this book — as well as the other books that I have completed — is part of why I am here.

In closing this section, I will add that, throughout my entire life and from my earliest recollections, I have always heard and seen things that did not exist to most other people. I also learned how to conduct *thought experiments* in the laboratory of my mind and to regularly present my findings to the Creator-God's Holy Spirit for approval as well as guidance and insights concerning refinement of my findings in subsequent investigations. All of my published books are products of my nurture, training, education, employment, Bible study, and Christian ministry in conjunction with my thought experiments.

An Introduction to Volume One

Here I am residing in Dayton, Tennessee. I am Joseph Adam Pearson, the furless and fearless anthropoid who is writing a book entitled *Intelligent Evolution*. It is more than coincidental that I am writing this book on harmonizing, or blending, the merits of creationism with the merits of evolution in Dayton, Tennessee because this city represents a metaphysical vortex in time and space where the two topics have already been brought together in a formal way. For the sake of clarification, *creationism* in this book is the doctrine that the Creator-God created everything in the physical universe as recounted in the first chapter of Genesis; and *evolution* in this book refers to: (1) cosmic evolution as interpreted by astronomical observations of the physically-observable universe; (2) biological evolution as interpreted by abiogenesis[1] and neo-Darwinism, the latter of which includes microevolution[2] and macroevolution;[3] and (3) consciousness evolution.

It was in Dayton, Tennessee that the Scopes Trial took place in 1925. *The State of Tennessee versus John Thomas Scopes* alleged that John Scopes, a high school science teacher, violated the Butler Act, a Tennessee state law passed in 1925 that forbad: (1) denying

---

1 ***Abiogenesis*** **describes the theory that the earliest life forms developed from inanimate matter in the primordial sea due to the unique conditions and circumstances present at the time.**

2 ***Microevolution*** **describes change of relative gene frequencies within a given species or one of its populations. Microevolution occurs over a relatively short period of time. The emergence of new species (speciation) may or may not occur at this level of evolution.**

3 ***Macroevolution*** **describes change that takes place above the level of species and occurs over a much longer period of time (i.e., at the level of geologic time scales). The emergence of new species (speciation) may or may not occur at this level of evolution.**

the Biblical account of creation; and (2) teaching the theory of evolution.

Although John Scopes was convicted of violating the Butler Act, his conviction was later overturned on the basis of a technicality. Because the Scopes Trial indulged both political and judicial monkey business, it did not provide for true academic debate between informed people from the two opposing sides. To be sure, "true academic debate" neither favors theology and philosophy over modern science nor vice versa. "True academic debate" simply allows for a rational discussion by people who have made themselves knowledgeable in relevant areas on both sides of the aisle that pertain to the topic at hand. For the sake of further clarification, "true academic debate" is not intended to be mere intellectual exercise; rather, it is intended to have practical applications and implications as well as broaden the insights and perspectives of those involved as players, participants, and evaluators.

Because the Butler Act was repealed in Tennessee in 1967, and because the Supreme Court overturned a similar state law from Arkansas in 1968, the prohibition of teaching evolution has been effectively overturned for public education in the United States. However, today, almost one hundred years after the Scopes Trial, many people still do not know how to effectively harmonize, meld, blend, or synthesize the various theological views on the Genesis account of Creation with the various modern scientific views on evolution. To be sure, many evangelical Christian schools, colleges, and universities — including Bryan[4] College in Dayton, Tennessee as recently as 2014 — are requiring their faculty to sign amended statements of faith that reinforce a rigidly narrow interpretation of the Genesis account of creation as well as repudiate interpretations to the contrary by branding any divergent thinking as non-

[4] **Bryan College was founded in 1930 and named after William Jennings Bryan, the prosecuting attorney for the State of Tennessee in the Scopes Trial.**

Christian. In doing so, these institutions are effectively removing academic theology from the domain of academic freedom. (Remember, in addition to providing the foundation for one's personal faith, theology is an academic content area that, ideally, should lend itself to passionate discussion, vigorous debate, and well-researched philosophical treatises).

One of the reasons that people fear academic debate comes from their own opinions that cloud their views as they navigate this life. The Satanic nature of Christian denominationalism prohibits Christians from thinking independently or, if they do, from opening their mouths to speak their minds — especially if they are members of an organized religious bureaucracy. Metaphysically, it is as if our opinions form electromagnetic force fields that prevent us, individually and collectively, from fully understanding topics about which we have preconceived notions and biases based on our own individual belief systems and limited intellectual conclusions (intellectual conclusions, including the conclusions of the present author, are always limited). One's personal *cloud of opinion* is impenetrable or penetrable based on the nature of the cloud and how the nature of each topic is perceived relative to that cloud. Individuals who have inquiring minds and permit their imaginations the necessary freedom to carefully consider the merits of new ideas, thoughts, concepts, and constructs have the least dense and, therefore, the most penetrable clouds of opinion. Thus, new truths are more readily available to such individuals because they are received in a positive way, regardless if they are ultimately accepted or rejected by those individuals. In contrast, new truths are obfuscated to those who permit their fears of the unknown to make their own individual clouds very dense and highly impenetrable. In an ideal world of education, training, nurture, and socialization, individuals are taught to think for themselves and not reject new ideas without first hearing them out and understanding their intended relevance and practicality as well as the posited reasons for their validity or invalidity.

This written work, *Intelligent Evolution*, is based on the following three major assumptions:

Assumption One

There is a personal Creator-God who is intimately involved in all events in the physical universe. This Creator-God is not simply the Prime Mover, First Cause, or Initiator of all events in the physical universe by, first, establishing laws associated with physics, chemistry, and biology and, then, allowing those laws to predetermine all subsequent actions, interactions, and reactions. Rather, this Creator-God is the living Supraconsciousness, or divine Mind, that continually provides the intelligent governing substrate for all events in the physical universe — past, present, and future. This Creator-God is neither physical nor housed in physicality. This Creator-God is invisible and indivisible. And this Creator-God is eternal. Despite being personal, the ways of this Creator-God are immeasurable and often beyond the comprehension of human beings. This Creator-God is the God of the Holy Bible, who is best experienced, understood, and known by human beings through their acceptance of His *only-begotten* Son, Christ Jesus, as Savior of the world and personal Savior.

Assumption Two

Metaphysics — the branch of philosophy and theology that includes the studies of *being* and *reality* (visible reality as well as invisible reality) — provides the best tool to bridge the gap of understanding: (1) between creationism and the "intelligent design" of all living things by a Creator-God; and (2) between evolution and its concepts of: (a) the cosmic evolution of the physically-observable universe; (b) the origins of all living things, such origins including their common ancestries and genetic variations resulting in, as well as explaining, the origin of different biological species and their adaptations to environmental change (collectively referred to in this book as *biological evolution*); and (c) consciousness evolution.

Christian metaphysics is the specific branch of theology that seeks to provide spiritual reasons for physical events; it remains *Christian* as long as it does not lose sight of the role of Christ Jesus as the only sacrificial atonement acceptable to the Creator-God for the iniquity and sin of souls in dust (i.e., corporeality), who actually became mortals because of their iniquity and sin. In this work, *Christian metaphysics* joins: (1) the theoretical to the empirical; and (2) the *unseen known* to the *seen known* in order to help bridge the gap between creationism and evolution. *Christian metaphysics* permits such synthesis because it gives us a major tool to understand the thinking of God.

Christian metaphysics is the best possible tool to help bridge the gap between creationism and evolution because it is capable of synthesizing and integrating the two views. In the book entitled *Intelligent Evolution,* the strengths of the present author are evidenced in his ability to perceive, apprehend, and think metaphysically at the same time that he tightly grasps the spiritual efficacy of the shed blood of Christ Jesus. Another practical ability of the present author for the writing of this book is his susceptibility, or sensitivity, to receiving words and images from otherworldly sources.

Major authors in the present author's studies of Christian metaphysics include: Aristotle (384-322 BC), Immanuel Kant (1724-1804 AD), Mary Baker Eddy (1821-1910 AD), Pierre Teilhard de Chardin (1881-1955 AD), and Stephen Hawking (1942-2018):

(1) Aristotle was a pantheist and philosopher. Although he was a pantheist, his contributions to Christian thinking, ethics, general philosophy, and metaphysics are immense. (2) Immanuel Kant was an agnostic and philosopher. Although an agnostic, his contributions to metaphysics are in providing a practical mental

framework for the ways in which intellectual knowledge is derived. (3) Mary Baker Eddy was a Christian Scientist and theologian. She was the first person in history to discover, develop, and refine a systematic theology based on Christian metaphysics. (4) Pierre Teilhard de Chardin was a Jesuit Priest and naturalist-philosopher. He was the first theologian with scientific training to provide Christian metaphysical explanations for biological evolution and consciousness evolution. And (5) Stephen Hawking was a theoretical physicist, cosmologist, and atheist who often speculated about the mind of God even though he could not believe that a personal God exists. He elaborated a cogent and convincing theory for cosmic evolution from the beginnings of the observable universe in its *Big Bang* through its current and continuing expansion as well as its requirement for the biological evolution of human consciousness.

Although the present author has formulated his own brand of Christian metaphysics, the development of his brand has been greatly influenced, not only by all five of the thinkers mentioned in the previous paragraph, but also by many other thinkers not specifically named in this book.

Assumption Three

The major difference between a contemporary understanding of creationism and a contemporary understanding of evolution is in the roles or non-roles of *chance* and *variables* in astronomical origins as well as the origins of species. Creationism, as espoused by most Christian fundamentalists, claims that there is no chance or variables in the origin of the cosmos and the various biological species. Evolution, as espoused by most modern scientists, claims that chance and variables were, and still are, in operation in the origin and maintenance of: (1) the physically-observable universe; (2) all extinct as well as currently existing biological species; and (3) all new species. However, rather than pitting creationism and the Creator-God's intelligent design of the physical universe against evolution and the roles of chance and variables in the ordered randomness of the physical universe, this work, entitled *Intelligent Evolution,* attributes all change in the physical universe — including all biological evolutionary change — to the spiritual means of the Creator-God, who provided, and still provides, teleological directionality to all cosmic, biological, and consciousness evolutionary changes at the same time that chance and variables are permitted to play roles in various aspects of those changes.

As used in this book, the noun *teleology* explains physical phenomena by the purposes they serve as determined by the Creator-God, who is the *First and Final Cause.* Thus, the adjective *teleological* includes starting from the end and reasoning backward (i.e., regressing) from the intelligent design of the physical universe to its *First and Final Cause.* Although traditional teleology only refers to the *Final Cause,* the present author has expanded it to include the *First Cause* in order to more accurately reflect teleology's link to the creative aspect of the Godhead in Christ Jesus, who is "the Alpha and the Omega," "the beginning and the ending," and the "first and the last" of all real *being (see Revelation 1:8, 1:11, 21:6, and 22:13).*

For the sake of additional clarification, *teleology* is the metaphysical explanation that the goals and intended results of the Creator-God, and not physical causes, guide design and purpose in the physically-observable universe, including all aspects of physical evolution. *Teleology* holds the doctrine that an intended outcome (that is, a results-driven, or purpose-oriented, cause) consciously guides all cosmic evolution, all biological evolution, and all consciousness evolution in the physically-observable universe. In this work, the *teleological* cause of the physical universe is not something nebulous but, instead, the opportunity for salvation through the shed blood of Christ Jesus extended to all salvageable fallen souls (which is to say, souls not beyond reclamation or redemption). The *teleological* cause of the physical universe is ultimately found in the Creator-God's Plan of Salvation, aptly and ably explained in the Holy Bible. The present author contends that the physically-observable universe exists as we know it solely for the purpose of our salvation and, thus, for the Glory of our Creator-God.

To be sure, the Creator-God is not adverse to chance and variables playing roles in evolutionary changes provided that they do not interfere with His Plan of Salvation through Christ Jesus. The Creator-God recognizes that the roles of chance and variables even augment His Plan of Salvation because they help to ensure healthy genetic diversity as well as biological success, succession, and ascendancy — all of which have augmented, and continue to augment, the final *teleological* cause of enhanced survivability, sustainability, and thrivability of one species, *Homo sapiens.* (Without the survivability, sustainability, and thrivability of *Homo sapiens,* opportunities for the salvation of fallen souls would be greatly diminished.)

The three major assumptions of *Intelligent Evolution* are explained more fully in the three major parts of this book, which address the following general topics in Volumes One and Two:

Part One
Creationism versus Evolution: Redefining the Problem

Part Two
Bridging the Gap between Creationism and Evolution:
Using the Tool of Metaphysics as a Problem-Solver

Part Three
The Theory of Intelligent Evolution:
Explaining the Solution to the Problem

Part One
Creationism versus Evolution: Redefining the Problem

1.1 Chance and Randomness

Chance is not as chancy as one might think. There is a science to chance. Chance is partly responsible for the ordered randomness in the physical universe, including the diversity of all living organisms. Because the outcomes of chance follow mathematical laws associated with probability (i.e., combinations and permutations), the outcomes of chance are predictable, especially with regard to dominant genes, co-dominant genes, recessive genes, multiple alleles (that is, sets of genes associated with individual traits or characteristics), genetic mutations (most of which are disadvantageous to a biological species, some of which are beneficial or, at least, non-harmful), and population genetics (most individual populations are in a dynamic equilibrium).

Our Creator-God employs chance to ensure that there is: (1) genetic diversity within each biological species, (2) genetic diversity within each ecosystem, (3) genetic diversity within each biome, (4) genetic diversity throughout the entire planet (specifically, in its *biosphere*), and (5) genetic diversity throughout the physical universe. (Did you really think that we are alone?)

Genetic diversity is ordained by our Creator-God because genetic diversity is healthy for each species, each ecosystem, each biome, and the planet's entire biosphere. Genetic diversity is healthy for each biological species because it ensures that an entire species is as capable as possible of surviving major changes to its ecosystem: *For example,* some individuals within each species will always be more fit than others in the same species to survive certain geologic and climatic changes. Genetic diversity is also healthy for the entire biosphere: *For example,* although large dinosaurs became extinct due to massive geologic and climatic changes throughout the entire globe during the Cretaceous-Paleogene extinction event approximately 66 million years ago, many of their smaller phylogenetic cousins did not become extinct, ensuring the survival of: (1) gene pools similar to the large dinosaur gene pools, and

(2) ecologic interrelationships similar to those involving large dinosaurs. And, because the teleological cause (that is, the first and final cause) for all alterations in gene pools is the eventual emergence, survival, and thrivability of *Homo sapiens,* the extinction of large dinosaurs — as well as approximately 75% of all plant and animal species on the planet Earth at that time — not only removed major threats to the future development of human society, culture, and civilization but also created evolutionary opportunities for the rapid advancements of then-existing species as well as the emergence of new species — all of which would ultimately pave the way for the eventual emergence of anthropoids and hominids (or *hominins,* depending on how up-to-date one's taxonomic terminology is).

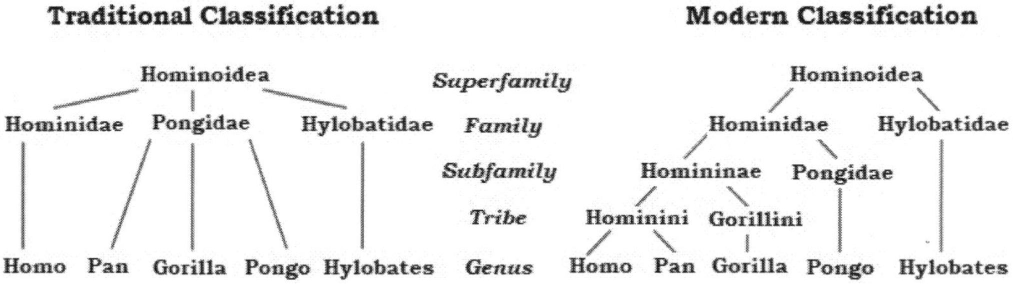

Note: In 2022, there is still debate if the genus *Pan* should be included within *Hominini.* (The genus *Pan* includes chimpanzees and bonobos.)

Is the present author implying that the Creator-God arranged for an asteroid to hit the planet Earth just at the right time for the purpose of drastically altering ecosystems to facilitate rapid evolutionary changes, the diversification of mammals, and the gradual emergence of anthropoids and hominids? The answer is "Yes."

Genetic diversity is sanctioned — that is, permitted, approved, and controlled — by our Creator-God to ensure that biological life on this planet *continues to continue* even though all individuals eventually die and some individual species eventually die out. Each biological species has been programmed by our Creator-God to be genetically diverse in variety, and all biological species have been purposely made by our Creator-God to be interdependent with other species. Our Creator-God ordained the role of chance in biological diversity in order to ensure the adaptability of each phylogenetic group to external changes, including competition for survival with organisms from its own group as well as with organisms from other groups. If some aspects of biological evolution are due to chance, then our Creator-God ordained it to be so. Nothing is done without our Creator-God's permission, approval, support, guidance, and direction.

Chance, as most people think of it in relation to natural selection, does not exist. It does not exist as most people think of it because biological diversity is nonrandom and the outcomes of mutations, adaptive radiations, and natural selection are predictable and, in most cases, inevitable. In other words, teleologically speaking, the origin of *Homo sapiens* was inevitable the first nanosecond that *the Big Bang* began (i.e., at the start of the inflationary epoch of the universe). Although individual events in the Earth's biosphere may seem to be random, it becomes apparent that they are nonrandom when events are grouped together in various sets throughout relative space-time. Here, the expression *relative space-time* means, physically speaking, that space cannot exist without time, and time cannot exist without space. In contrast, metaphysically speaking, absolute time *is* absolute space.

Chance under the auspices of the Creator-God is *directed chance* and not haphazard, or uncontrolled, chance. *Directed chance* played an important role in creation-evolution, and *directed chance* plays an important role in the eventual extinction of human somatic identities and their replacement by spiritualized somatic identities (i.e., *astral gelatinous™* forms for saved people). Indeed, in the not-too-distant future, all physical forms will cease to exist.

Note: The phrase *astral gelatinous™* was coined by the present author and first copyrighted in the 2011 edition of his work entitled *Divine Metaphysics of Human Anatomy* (United States Copyright Office TXu001788674). Simply stated, the phrase represents spiritual living substance, which, if seen by human beings, would appear semisolid and translucent.

Although an understanding of random genetic mutations is an integral part of contemporary neo-Darwinism, genetic mutation was not a part of classical Darwinism in 1859, the year that Charles Darwin first published *On the Origin of Species* — or, more completely, *On the Origin of Species by Means of Natural Selection, or the Preservation of Favoured Races in the Struggle for Life*. (As a side note, Darwin did not use the word *evolution* in *Origin of Species* until its 6th edition in 1872.) To be sure, Charles Darwin elaborated on the roles of variation and selection in the struggle for existence and the survival of the fittest. But he was not aware of: (1) genes determining inheritance, (2) principles of molecular genetics, including deoxyribonucleic acid (DNA) and ribonucleic acid (RNA), or (3) random genetic mutations.

Just as chance is not as chancy as one might think, new successful, niche-filling species variations are not as random as one might think nor is adaptation at the organismic (organismal) level as complex as one might think. Evolution is actually based on a limited number of genes mutating. *For example,* the primary morphological differences between a hummingbird and an ostrich are in their skeletal systems. Why *primary?* Structural bone changes from beneficial or non-harmful inherited genetic mutations are accompanied by epidermal, nerve, muscle, and vessel accommodative processes based on the adaptability of epidermal, nerve, muscle, and vessel precursor cells. Co-evolution of the epidermal cells, nerve cells, muscle cells, and vessel cells at the genetic level is not required to accompany gross skeletal changes: Epidermal cells, nerve cells, muscle cells, and vessel precursor cells simply migrate into body tissues that require protective covering, innervation, movement, and oxygenation without requiring unique heritable genetic mutations in the precursor cells. (The present author is not suggesting that ostriches evolved from morphological changes in hummingbirds.)

Overall, mechanisms for generating novelty in evolution are dependent on: (1) genetic mutations, (2) changing environmental conditions, and (3) organismic exploratory processes through *directed chance* and *facilitated variation*. The present author's metaphysically-nuanced meanings for *directed chance* and *facilitated variation* include the growth of cells, tissues, organs, organ systems, organisms, and biomes into ideas that already exist in the Mind of the Creator-God and opportunities that already exist in nature.

Note: The present author views adaptation at the organismic level as the fulfillment of potential that has both supernatural and natural components.

Relative to chance and randomness, there are only two possible perspectives for understanding evolution in an intelligent and intelligible way:

1. Evolution through the lens of scientific atheism assumes that a purposeless physically-observable universe created itself over billions of years through self-organizing emergent systems resulting in cosmic evolution, biological evolution, and consciousness evolution. This perspective posits that unusual patterns subsumed in chaos emerge as increasingly complex organized systems through physically-sparked alterations of expected outcomes from chance and randomization.

2. Evolution through the lens of Christian metaphysics assumes that a purpose-driven physically-observable universe was created by the God of the Holy Bible over billions of years through His actualizing organized emergent systems resulting in cosmic evolution, biological evolution, and consciousness evolution. This perspective posits that unusual patterns emerge from chaos as increasingly complex organized systems in outcomes of God's directed chance and facilitated variation.

Not surprisingly, the present author's paradigm of *intelligent evolution* supports the perspective of Christian metaphysics.

1.2 The Whole Universe

In this book, *the Whole Universe* is divided into: (1) the spiritually- or metaphysically-observable universe; (2) the physically-observable universe; and (3) the empty vacuum of space beyond the fringes of the physically-observable universe. (The fringes of the physically-observable universe constitute the *cosmic horizon*.) The physically-observable universe and the empty vacuum of space beyond its fringes together comprise the entire physically-knowable universe — which is largely *knowable* through the physical senses and instrumentation. (*Instrumentation* includes mechanical and technological extensions of our physical senses.) In contrast, the spiritually- or metaphysically-observable universe constitutes the entire unknowable universe — which is largely *unknowable* because it is unknowable to the corporeal, or physical, senses and instrumentation — although it is, of course, knowable to the spiritual, or metaphysical, senses through the Creator-God's Holy Spirit.

In his other books on Christian metaphysics, the present author uses the phrases *material universe* and *physical universe* interchangeably. However, in this book, the phrase *material universe* is not synonymous with the phrase *physically-knowable universe* or the phrase *physically-observable universe*. Rather, here, *material universe* refers more specifically to the universe of ordinary matter (elemental, atomic, or baryonic matter) — which is to say, in this book, the phrase *material universe* is not meant to include either dark energy or dark matter. Certainly, an understanding of ordinary matter as the *material universe* is consistent with what Aristotle, Kant, Eddy, and even de Chardin knew of the entire universe. To them, the phrase *material universe* sufficiently described the physical universe as they understood it. Among other things, they had never heard of dark energy nor of dark matter; simply stated, robust evidence for the existence of dark energy and dark matter was not widely reported until after their lifetimes. This, of course, was not true for Stephen Hawking.

Incorporating principles of physics and metaphysics, following are seventeen foundational axioms that the present author has used to conceptualize the major components of *the Whole Universe:*

1. Matter is the substance, or essence, of the physically-observable universe.

2. Matter is anything that has mass and takes up space.

3. Mass is a fundamental property of matter.

4. Mass is one way that matter can be measured.

5. Using mass-energy equivalents is a specific way to measure matter that takes into account the interconvertibility of mass and energy in the physically-observable universe. (See the "mass-energy equation" immediately following this list of seventeen axioms.)

6. If mass is a fundamental property of matter, then matter is a function, or root, of mass.

7. For the physically-observable universe, mass and physical energy are properties of consciousness but consciousness is not equivalent to matter.

8. If mass and physical energy are properties of consciousness, then, consciousness is a function, or root, of mass and physical energy in the physically-observable universe but not in the spiritually-observable universe because there is no mass or physical energy in the spiritually-observable universe.

9. Spirit is the substance, or essence, of the spiritually-observable universe; and spiritual energy is the causative agent in the spiritually-observable universe. (The Creator-God is *the First and Final Cause* of *the Whole Universe* and its component parts.)

10. Spirit is massless and takes up no space in the spiritually-observable universe because there is no space in the spiritually-observable universe (as human beings understand space).

11. Spirit and spiritual energy are ways that consciousness can be measured in the physically-observable universe but not in the spiritually-observable universe because Spirit and spiritual energy are immeasurable in eternity.

12. Spirit and spiritual energy are fundamental properties of consciousness.

13. Spirit and spiritual energy are fundamental properties of consciousness in both the physically-observable universe and the spiritually-observable universe.

14. If Spirit and spiritual energy are fundamental properties of consciousness, then consciousness is a function, or root, of both Spirit and spiritual energy as well as both mass and physical energy.

15. Consciousness is a function, or root, of Spirit and spiritual energy in the physically-observable universe as well as in the spiritually-observable universe.

16. Given conditions established, required, and fulfilled by the Creator-God, spiritual energy and physical energy are inter-convertible within *the Whole Universe*.

17. The Creator-God is also the Creator-Evolver in addition to the Creator-Savior.

Based on the interconvertibility of mass and energy ($E = mc^2$ or $m = E/c^2$), the mass-energy, or mass-equivalent, content (μ) of the physically-observable universe is represented by the equation that follows:

$$\mu_{de} + \mu_{dm} + \mu_{H+He} + \underbrace{(\mu_{eh} + \mu_{am} + \mu_{ee})}_{\text{negligible}} = 100\%$$

68.3% = μ_{de} = dark energy (intrinsic energy of vacuum space within the physically observable universe)

26.8% = μ_{dm} = dark matter

4.9% = μ_{H+He} = hydrogen (H) and helium (He)

<.1% {
μ_{eh} = elements of ordinary matter heavier than H and He
μ_{am} = antimatter
μ_{ee} = everything else

In the physically-observable universe, the following constitute such a small percentage that, even altogether, they are fractionally negligible — or effectively zero — in relation to the whole: (1) the mass-energy content of elements of ordinary matter heavier than hydrogen and helium; (2) the mass-energy content of antimatter *(for example,* positrons, antiprotons, and antineutrons); and (3) the mass-energy content of everything else *(for example,* electro-magnetic radiation). That they are fractionally negligible does not mean that they have no value. *For example,* illustrating the Creator-God's propensity for creating something out of nothing as well as making something out of next-to-nothing, it is primarily from the mass-energy content of elements heavier than hydrogen and helium that the Creator-God spoke into existence our solar system, including our Earth and all biological life on it. (1) Because more than 99% of the bulk composition of the Earth (by elemental-mass) is composed of elements heavier than hydrogen and helium, and (2) because more than 90% of all living substance is composed of elements heavier than hydrogen and helium, our Creator-God, once again, has illustrated that He takes from what is rare to form what is precious in order to further magnify and glorify His Holy Name — in this case, through the creation-evolution of the cosmos and biological life.

The truth be told (and it is being told right here), the physically-observable universe is still evolving. Supernovae (supernovas) still produce every element of ordinary matter possible: And all elements heavier than hydrogen and helium are still created by (1) *fusion* in the combination of various hydrogen and helium atomic nuclei and (2) *spallation* in the ripping apart of atomic nuclei of heavier elements and reconfiguring them into atomic nuclei of lighter elements (a kind of fission-fusion, one might say). Indeed, the entire physically-observable universe is still in the process of being created. This ongoing creation-evolution has helped the present author conclude that the Genesis account of cosmology — as well as of abiogenesis and biogenesis on Earth — is primarily an account of the creation and evolution of our individual planet in relation to the rest of the physically-observable universe.

At this juncture, it is important to emphasize that the physically-observable universe is finite for two major reasons:

1. *The physically-observable universe exists only for a finite time.* It began with *the Big Bang,* and it will be dissolved at the end of the millennial rule of Christ Jesus on Earth. Its finite age is even attested to in the Bible by: (a) the Genesis account of creation; (b) its presentation by God the Son to God the Father at the end of his millennial rule *(1 Corinthians 15:24-28);* and (c) the creation of "a new heaven and a new earth" *(Revelation 21:1).*

2. *The physically-observable universe occupies only a finite place.* It ends in space at its fringes, or cosmic horizon.

The physically-observable universe is not eternal and was never meant to be eternal — nor can it ever become eternal or ever be our eternal, or heavenly, home. Only the spiritually-observable universe is eternal. In this book, the terms *infinite, infinitude, infinity, finite, finitude,* and *finity* do not apply to the spiritually-observable universe but the terms *eternal* and *eternity* do. In contrast to the spiritually-observable universe, the entire physically-knowable universe is infinite because there is no physical end to the empty vacuum of space beyond the physically-observable universe, but the physically-observable universe itself is finite because there is a physical end to it. (Technically, because the physically-observable universe is ever-expanding, it has no outer physical barrier or boundary.) (See *Note* on page II-77.)

Metaphysically speaking, even the cosmic infinite has a beginning and an ending: the nothingness of infinity began at the time of *the Big Bang,* and it will end when the physically-knowable universe ceases to exist. For the sake of clarity, when used by the present author, the terms *eternal* and *eternity* never apply to the physically-knowable universe — neither its physically-observable universe nor the infinite and empty vacuum of space beyond its fringes.

The finite has a beginning and an ending. In contrast, eternity has neither a beginning nor an ending. Because the Creator-God is eternal, the Creator-God is neither infinite nor finite. Metaphysically speaking, the Creator-God is neither too big nor too small. Because the Creator-God is eternal, the Creator-God is ageless, dimensionless, and motionless: Our Creator-God does not have age, dimensionality, and movement as corporeal beings understand age, dimensionality, and movement. In this book, the words *eternal* and *everlasting* are synonymous; and the word *forever* refers to the full duration of relative time in the physically-observable universe. In other words, *eternal* and *everlasting* are not synonymous with *forever.* In this book, *forever* lasts only up to the time of the creation of "a new heaven and a new earth" *(Revelation 21:1 KJV)* at the end of the millennial reign of Christ Jesus.

Although the term *universe* is used in multiple ways in this book, its plural uses are for the two major components of *the Whole Universe* and not meant to convey that *the Whole Universe* includes a multiverse of multiple physical universes. And, although the physically-observable universe and the empty vacuum of space that surrounds it will both disappear one day (that is, cease to exist because they will be transmogrified to, translated into, infused by, returned to, and swallowed up by the substance of the Creator-God's Holy Spirit), *the Whole Universe* will still remain, but it will then be composed only of what is referred to in this book as the spiritually- or metaphysically-observable universe.

The spiritually- or metaphysically-observable universe (also known in this book as *the Spiritual Universe, Heaven, the Creator-God's Heavenly Consciousness, the Supraconsciousness of the Creator-God, Paradise,* and *Eden*) is spiritual, immortal, and eternal. In contrast, the physically-observable universe is physical, mortal, and finite. Again, although the empty vacuum of space that surrounds it is infinite, the physically-observable universe itself is finite because there is an end to it at its fringes (indeed, as mentioned earlier: although it has no physical boundary, it does have a temporal end when it ceases to exist). The spiritually- or metaphysically-observable universe is the state and place where the Creator-God's Holy Spirit is substance. Correctly apprehended by Aristotle, the substance of a thing is its essence and, conversely, the essence of a thing is its substance. Therefore, Spirit constitutes both the substance and essence of spiritual things in the spiritually- or metaphysically-observable universe; and matter constitutes both the substance and essence of physical things in the physically-observable universe. Spirit is the primary reality of the Creator-God. Matter is only a secondary reality.

The spiritually- or metaphysically-observable universe is real. The physically-observable universe is also real but in a different way. Some metaphysicians have tried to pit Spirit against matter when they should have been pitting Spirit against Evil. Although realities of the spiritually-observable universe and the physically-observable universe may intersect and interact at times, they are largely separate from one another. Metaphysically speaking, the two universes are conjoined at the same time that they exist in tandem. (See Figure One.)

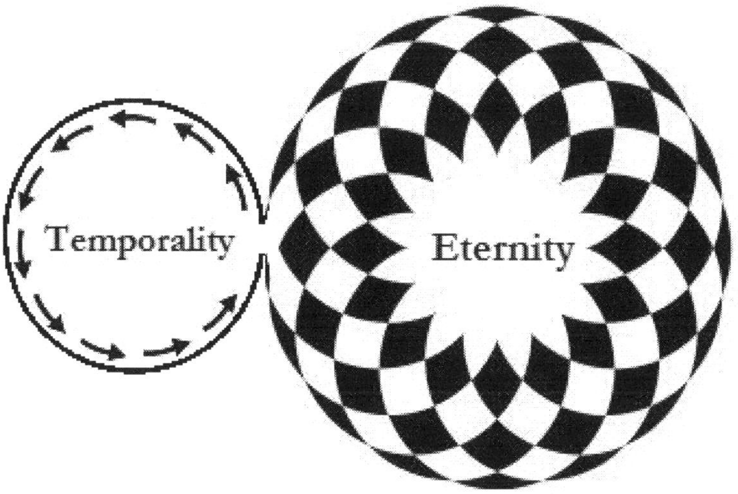

The Outpocketing of Temporality from Eternity

Figure One

The spiritually-observable universe and the physically-observable universe represent two separate creations, or elaborations, of the Creator-God. In effect, the physically-observable universe was ordained by the Creator-God in order that spiritual beings who *fell* from the spiritually-observable universe would have something practical to fall into. Today, the primary focus of the Creator-God throughout *the Whole Universe* is the salvation of fallen, or errant, souls who were immortal before their fall but are now mortal due to their fall. The Creator-God uses corporeality to achieve this end as it relates to His Plan of Salvation for souls who inhabit the corporeal bodies of *Homines sapientes* (the plural of *Homo sapiens*).

Although Adam and Eve were originally created as spiritual beings, they fell from their first estate in the spiritually-observable universe.

When they fell, they *fell* by the design of the Creator-God into anthropoid bodies belonging to the genus and species of *Homo sapiens* on the planet Earth. In other words, the fallen Adam and Eve materialized as human beings alongside of other similar hominins that had already evolved on the planet Earth. (Here, *similar* is referring specifically to hominins with 46 chromosomes.)

In the appearance of hominins on the planet Earth, the Creator-God used biological evolution to create an anthropoid species capable in complexity of eventually housing the fallen souls of spiritual beings. Here, *capable in complexity* is especially referring to a central nervous system with highly developed cerebral hemispheres, including frontal lobes sufficiently able to permit higher order thinking, memory, and imagination.

In other words, when the fallen spirit beings of Adam and Eve materialized as two human beings, other hominins were already living at that time, but they were soulless. Thus, when Cain, the banished son of Adam and Eve, went to live with the people of Nod *(Genesis 4:16-24),* he was with soulless hominins who were living in an adjacent region east of the portal through which Adam and Eve were expelled from the Garden of Eden *(Genesis 3:23-24)*.

Eve is called "the mother of all living" *(Genesis 3:20)* not because she was the mother of all *H. sapientes* but because she is the mother of all hominins who have souls. Indeed, all descendants of Adam and Eve are *H. sapientes* with eternal souls. Our individual fallen souls are temporarily fused, or tethered, to individual human bodies in order for us to learn to grow back to the Creator-God as well as for us to have opportunities for eternal salvation — that is, the restoration of our fallen souls to immortality — by embracing the shed blood of Christ Jesus as the only sacrifice acceptable to the Creator-God for the forgiveness of our iniquity and sin.

Except for Adam and Eve, no other hominins living at the time that Adam and Eve fell to Earth possessed eternal souls. Thus, the people of Nod did not possess eternal souls but the children of

Cain did because they were direct descendants of Adam and Eve through Cain. Thus, all members of *H. sapiens* today have souls because they are all direct descendants of Adam and Eve. There are no soulless hominins living today because none of them survived the cataclysmic flood that occurred during Noah's time.

If the author of *Intelligent Evolution* were responsible for the taxonomic nomenclature describing these two early groups of anthropoids, he would name hominins without souls *Homo sapiens* var. *sine anima*[5] and those with souls *Homo sapiens* var. *cum anima*.[6]

All order in the spiritually- or metaphysically-observable universe has been propagated and maintained by the Creator-God. All cosmic and biological order in the physically-observable universe has also been propagated and maintained by the Creator-God. Any and all order and non-randomness (that is, negative entropy, or *negentropy*[7]) that exists in the physically-observable universe is a reflection of the order that exists in the spiritually- or metaphysically-observable universe. And any and all disorder and randomness (that is, entropy) in the physically-observable universe is a direct result of the Luciferian Fall.

Failure to recognize and apprehend the significance of the Luciferian Fall results in one's failure to understand the desirability for the return of fallen souls to the spiritually- or metaphysically-

5 "*Homo sapiens* var. *sine anima*" refers to the "modern man variety without a soul."

6 "*Homo sapiens* var. *cum anima*" refers to the "modern man variety with a soul."

7 *Entropy* describes "the loss of usable energy in a system and the measurement of that system's change from order to disorder." In contrast, *negentropy* describes "the gain of usable energy in a system and the measurement of that system's change from disorder to order."

observable universe. And, for the sake of comparison and contrast, there is no entropy (i.e., loss of energy) in the spiritually- or metaphysically-observable universe: Divine substance and divine energy are never lost in Spirit although some divine substance and divine energy were altered by being converted into physical substance, or matter, and physical energy at the time of the Luciferian Fall. (The Luciferian Fall is metaphysically coincident with *the Big Bang*.) In the spiritually-observable universe, *divine substance* and *divine energy* are never diminished because they possess the unique trait of self-propagation.

"Divine substance" is Spirit (i.e., the Creator-God's Holy Spirit) and "divine energy" is the *eternal energy,* or *divine fire,* of the Godhead. In the Greek New Testament, *Theos* (θεός) means "the supreme Divinity;" *Theios/Theiotes* (θεῖος/θειότης) means "Godhead;" and *Theion* (θεῖον) means "divine fire," which is "the *eternal energy* of the Creator-God." In this book, the anglicized word *theion* (the English plural form is *theions*) provides a useful neologism. A *neologism* is "a newly-devised word or a new sense to an already existing word." For the purpose of *Intelligent Evolution,* a *theion* is "the smallest indivisible unit of divine, or eternal, energy." (This definition satisfies the "new sense" aspect of a *neologism*.) The following metaphysical analogy (or parallelism) might help the reader or listener to understand: "*theion* is to divine energy and divine light as *photon* is to physical energy and physical light." Just as a *photon* is a force-carrying, massless particle in the physically-observable universe, so is a *theion* a force-carrying, massless particle in the spiritually-observable universe.

One measure of the utility of the *photon* to *theion* comparison arises in the capacity of the units to self-propagate or not. Because photons are not able to self-propagate and *theions* are able to self-propagate, the *photon* to *theion* comparison is less than perfect. However, it is still a useful analogy, and conceptualizing *theions* provides a useful paradigm for understanding divine substance and divine energy in the spiritually-observable universe.

For the sake of clarification, the reason that *theions* are able to self-propagate is that they are composed of divine love in addition to divine light. (Indeed, divine light and divine love are inseparable and are only mentioned here separately for the sake of discussion.) The Creator-God Himself is composed of *theions*. Thus, the Creator-God's very nature, or essence, includes His desire to self-propagate — or, in this case, to make created beings in His complete image and perfect likeness. To be sure, this desire is born of His divine love. His divine love wants (no, *needs*) to be shared with others in fellowship, communication, compassion, tenderness, mercy, grace, and care. Because the Creator-God *is* divine Love *(1 John 4:8 and 4:16),* He wants (no, *needs*) to share the largess of it with beings created in His complete image and perfect likeness.

The only danger in understanding the paradigm of *theions* is in the misguided conclusion that one can know the unknowable or can reduce the omnipotent, omniscient, and omnipresent Creator-God to one's own terms of understanding. What guards against operating in this misguided conclusion is one's ability to live in a state of perpetual contrition, which state is against the fallen nature of being human but very much a part of the unfallen nature of being divine — that is, a part of the spiritual creation of the Creator-God — which includes being recast in the spiritual image and likeness of the Creator-God through the shed blood of Christ Jesus.

Any and all order and non-randomness that exists in the physically-observable universe becomes understandable to us when we understand the Creator-God and His deific Force, which Force is His spoken Word, creative Logos, or divine Principle — by which He creates, operates, gathers, and restores. All of the laws that provide the governing substrate of *the Whole Universe* are elaborated by the Supraconsciousness, or divine Mind, of the Creator-God. The creative Logos, or divine Principle, of the Creator-God permits human beings to again possess the Earth (i.e., have dominion over it) by overcoming Evil, iniquity, and sin through the shed blood of Christ Jesus. "Mortality is swallowed up by immortality" *(2 Corinthians 5:4)* only through the metaphysical application of that blood. Although Christians may not be able to stop an active volcano from killing them, they can stop an active volcano from impacting negatively on the immortal life that has been restored to them through Christ Jesus.

Although one should aim for a literal understanding concerning the shed blood of Christ Jesus, the application of the shed blood of Christ Jesus to earthborn problems is metaphysical — which application is, in one way, neither literal nor figurative and, in another way, both literal and figurative. Applying the shed blood of Christ Jesus to earthborn problems requires us to metaphysically look at all problems through that blood. It is through such a view that earthborn problems become *resolved* — meaning, understood as well as solved. The resolution of an earthborn problem is: (1) always metaphysical primarily; and (2) only physical secondarily, if at all.

Partitioning our thinking is necessary to understand both the spiritually-observable universe and the physically-observable universe at the same time in order that we might have a metaphysically stereoscopic view of *the Whole Universe*.

To be sure, the physically-observable universe is not only a metaphysical allegory of the spiritually-observable universe but also an inverted reflection of the spiritually-observable universe. With that said, however, the physically-observable universe is not a parody, a perversion, or an imitation of the spiritually-observable universe; it is simply the Creator-God's second creation, or second elaboration. All order in the physically-observable universe is a metaphysical representation of the order in the spiritually-observable universe. Although physics may seem to govern the physically-observable universe, metaphysics actually does. And cosmic, biological, and consciousness evolution reflects the spiritual evolution, spiritual phylogeny, and spiritual ontogeny that occurred, and still occurs, in the first creation of the Creator-God, the spiritually-observable universe.

For the sake of clarity here, evolution in the spiritually-observable universe is not like evolution in the physically-observable universe. Entities in the spiritually-observable universe do not evolve into new entities; and new species do not arise in the spiritually-observable universe. Instead, it is more on point to say that the spiritually-observable universe continues to expand in the Supraconsciousness of the Creator-God. This spiritual expansion constitutes spiritual evolution.

Because the Supraconsciousness of the Creator-God continues to expand, so does the consciousness of His entire creation, including His spiritually-observable creation as well as His physically-observable creation. Thus, *the Whole Universe* continues to evolve and expand.

Like the consciousness of His entire creation, the Supraconsciousness of the Creator-God expands and will continue to expand. *For example,* before the Creator-God came to Earth as God the Son, the Creator-God had never experienced temptation for Himself *(James 1:13)*. Because the Creator-God is omniscient, He knew what temptation is and could have dictated a highly accurate 100,000 volume encyclopedia about it. But the Creator-God's knowledge of temptation was only academic — which is to say, His knowledge of temptation was not experiential (i.e., personal and intimate by having been tempted Himself). However, through the experiences of Christ Jesus, the Creator-God's knowledge of temptation is now not only academic but also experiential. What God the Son learned about temptation *(Hebrews 4:15)* while he was on Earth was shared synchronously and simultaneously with the rest of the Godhead. (For the sake of clarity, the threefold Godhead consists of God the Father, God the Son, and God the Holy Spirit.)

The Creator-God evolves Himself by expanding His Supraconsciousness. And the Creator-God evolves His creation by expanding the consciousness of His created beings. The Creator-God even evolves human beings by expanding their understanding of Him. By permitting us to experience and overcome Evil for ourselves, the Creator-God has brought us closer to His divine level of knowledge and understanding. Although we can never become the Creator-God, we can become more like Him and, thereby, make a more suitable eternal companion for Him — individually, collectively, and corporately.

By permitting Himself to experience temptation through the life experiences of God the Son, God the Father has also brought Himself closer to fallen created beings. Through the life experiences of Christ Jesus, the entire Godhead now knows experientially what it means to be vulnerable to temptation when one's soul is in human flesh. As God in the flesh *(1 Timothy 3:16 KJV),* Christ Jesus was touched with and by our infirmities. God's eternal mercy flows to us, first and foremost, through the shed blood of His *only-begotten* Son, but it is also effluent because of the Creator-God's firsthand understanding of our condition in corporeality through the earthly experiences of His *only-begotten* Son. The entire Godhead has experienced temptation, victimization, and the shedding of innocent blood personally through God the Son.

That the Supraconsciousness of the Creator-God expands and will continue to expand is not in conflict with the truth that God never changes. To be sure, the Creator-God's substance and nature never change. The Creator-God's personal species, kind, substance, and essence never change. But He continues to consciously expand the substance and nature of His Being. In addition, the Creator-God is ever-expanding experientially. If the reader or listener thinks about it, this is what one should expect from the Godhead because the Creator-God is dynamic and not static. The Creator-God's divine and universal Mind remains insatiably inquisitive and curious at the same time that it is creative. The Creator-God continues to create and expand Himself into His ever-expanding spiritually-observable universe. The totality of an ever-expanding Creator-God can only fit into the totality of an ever-expanding Creation. (In this way, there is a parallel, or analogy, between the infinite vacuum in which the physically-observable universe is located and the eternity that encompasses the spiritually-observable universe.)

For readers or listeners who may have taken offense on behalf of the Creator-God: (1) because the present author has stated that the Creator-God is "insatiably inquisitive and curious," and (2) because they feel that this statement is inconsistent with the Creator-God's omniscience, please know that the Creator-God endowed created beings with free will so that He might interact *with* them as well as be challenged *by* them. The Creator-God is not content with just observing His created beings; the Creator-God wants (no, *needs*) to interact with us. To be sure, the Creator-God wanted, and still wants, an eternal companion in us all individually, collectively, and corporately, but the Creator-God does not want His eternal companion to be composed of predictable and robotic automatons. It pleases the Creator-God to interact with our own creativity-in-action, especially as it is intended to honor Him by reflecting His complete image and perfect likeness. To be sure, the Creator-God is the source of our individual, collective, and corporate creativity — including our creative imaginations.

For the sake of clarity, the Creator-God does not devolve. Unlike the free will members of His original creation (that is, immortal beings who became mortal beings through their own iniquity and sin), the Creator-God cannot devolve. Devolution can only occur in segments, aspects, and parts of the Creator-God's creation when created beings consciously choose to depart from Him by stepping outside of His Will through disobedience. Of course, this happened to Lucifer and the angels who fell with him as well as to Adam and Eve; and it still happens to the souls of mortal beings who consciously (that is, *willfully*) reject the Creator-God by rejecting His Plan of Salvation and, thereby, continue to disobey His Supreme and Sovereign Will. As a result of their irrevocable rejection of the Creator-God, the souls of all eternally-reprobate mortal beings become the demons, devils, evil spirits, and unclean spirits described in the Holy Bible (all four terms are used synonymously throughout this book as well as within various translations and versions of the Holy Bible).

In His omnipotence, the Creator-God has permitted Satan, demonic forces, Evil, iniquity, and sin to exist but only for a predetermined time. Satan, demonic forces, Evil, iniquity, and sin — all of which constitute *spiritual chaos* — will be expunged at the end of the millennial rule of Christ Jesus on Earth. Unfortunately, some people erroneously presuppose that the Creator-God is already "All-in-all" everywhere. However, the Holy Bible is clear that the Creator-God, who is "All," only becomes "All-in-all" after the millennial reign of Christ Jesus on Earth has ended *(see 1 Corinthians 15:28 KJV),* when God the Son hands everything over to God the Father. "True man," "Man," "immortal man," or "original Man" (all synonyms here for the unfallen, immortal beings collectively known as *Adam*) was first created spiritually; it was only when Adam permitted self-will and self-pride to take hold that Adam fell from immortality to mortality. Fortunately, it is through the shed blood of Christ Jesus that the souls of "fallen man," "mortal man," or "the lost Adam" are fully restored to Spirit. (Efforts to restore lost souls to immortality without their accepting the shed blood of Christ Jesus as atonement for their iniquity and sin are of no avail.) The final translation of the physical creation back into Spirit at the end of *the Millennium* requires the full metaphysical application of the shed blood of Christ Jesus to everything restorable, reclaimable, and redeemable that has been outside of the spiritually-observable universe in temporality since the Luciferian and Adamic Falls.

The Whole Universe currently contains tandem creations, or two elaborations: one spiritual and one physical. The physical was created, or manifested, to *catch* eternal souls when they fell. (The Adamic Fall and the expulsion of Adam and Eve from Eden were synchronous.) Depending on where you are standing relative to eternity, it can appear to you that the substance, or essence, of the spiritually- or metaphysically-observable universe was altered as it fell to become the substance, or essence, of the physically-observable universe. For that reason, the entire physically-knowable universe might also be referred to as *the altered universe*. Regardless, it is important to conceptualize the Creator-God's tandem creations not only as conjoined universes but also as overlying parallel universes that are inverted reflections of one another.

Human beings live in a metaphysically rotated, or refracted, version of the spiritually-observable universe. That is why, when we are in the right frame of mind, we can catch an inward glimpse of the spiritually-observable universe now and then. The substance, or essence, of the spiritually-observable universe is Spirit. And the substance, or essence, of the entire physically-observable universe is matter and the physical energy into which matter's mass can be converted and vice versa. (See the earlier discussion on the mass-energy content of the physically-observable universe in Section 1.2.) Serious students of Christian metaphysics need to be reminded frequently that it is a huge mistake for them to pit Spirit against matter because they will be fighting the wrong enemy. Instead, they need to pit Spirit against its true enemy, Evil. Human beings do not need to forsake matter, but, instead, they need to forsake Evil by overcoming iniquity and sin. All redeemed beings in Christ Jesus, including those who currently reside in human bodies as well as those who currently live in spiritual bodies, are comfortable and satisfied no matter where they are. Human beings do not need to deny the existence of matter or physical conditions. Believing that something does not exist when it does exist is accompanied by unnecessary difficulties as well as ongoing consternation.

The spiritually- or metaphysically-observable universe is eternal, and the created beings housed in it are immortal, not mortal. Because the souls of human beings move from mortality to immortality through the shed blood of Christ Jesus, the teleology of Teilhard de Chardin is not so far afield when we understand that cosmic evolution, biological evolution, and consciousness evolution are all purpose-oriented, moving in the direction of a greater complexity that is more reflective of true spiritual being.

Where de Chardin misses the mark is in his understanding of the *Omega Point.* The entire physically-knowable universe is not rushing to become the spiritually- or metaphysically-observable universe because, at the end of *the Millennium,* the entire physically-knowable universe will be resorbed by, translated into, and infused, or swallowed up, by the spiritually-observable universe. At that time, the currently-existing tandem creations, or two parallel universes, will again become one.

The Whole Universe, including the spiritually-observable universe and the physically-observable universe, is the manifestation of the Creator-God's deific Force. Created beings who live in the spiritually-observable universe are real. And created beings who live in the physically-observable universe are also real. Human beings are merely God's created beings in His second estate.

Both immortal man and mortal man each have their own realities: These two groups of created beings each have a different referent and possess a different substance, or essence, in their respective realities. However, one is no less real than the other even though each group is in a different state and condition of being. (For the sake of clarity, *mortality* and *immortality* are states of being, and *corporeality* and *incorporeality* are conditions of being: Some *immortals* are corporeal in that they are saved fallen souls who have not yet returned to Heaven; and some *immortals* are incorporeal in that they have already returned to Heaven. All *mortals* are unsaved fallen souls. Depending on where they are located in *mortality,* some *mortals* are incorporeal, and other *mortals* are corporeal.)

As introduced earlier, evolution is a process that occurs in both the spiritually-observable universe and the physically-observable universe. It is not that the Creator-God improves upon His work in the spiritually-observable universe — instead, He simply expands upon it. Just as the Creator-God is not stagnant, so are His created beings not stagnant. The Creator-God co-exists with His created beings as well as inhabits them. Although the Creator-God governs *the Whole Universe,* He only inhabits the spiritually- or metaphysically-observable universe and saved souls in corporeality. The physically-observable universe cannot hold the spoken Word, divine Principle, or creative Logos of the Creator-God, except: (1) in a metaphysical sense, (2) in the singular instance when the Creator-God took on flesh as Christ Jesus, and (3) in saved souls who remain in corporeality because their human life spans have not yet ended.

Because the Creator God is eternal, He does not have a beginning or an ending. Likewise, because His first creation, the spiritually-observable universe, is eternal, it also does not have a beginning or an ending. In contrast, because his second creation, the physically-observable universe, is not eternal, it does have a beginning and an ending. When the present author refers to the *infinite* axes of eternity elsewhere in his literary works, it is a figurative reference and not a literal one: The spiritually-observable universe possesses only metaphysical axes. Although the physically-observable universe has a center and circumference physically, the spiritually-observable universe only has a center and circumference metaphysically. (Our Creator-God is both the center and the circumference of the spiritually-observable universe.)

All souls were created in eternity before the beginning of the physically-observable universe. All souls were created at the same instant in eternity through the same vocalization, articulation, and actualization of the Creator-God. Because souls were *created,* all souls have a beginning. However, their common beginning cannot be understood in terms of chronological time because all souls were created *in eternity*. It may sound strange to the reader or listener, but, once souls were created, it was as if they always *were* (just as they always *are* and always *will be*). Because all souls live, move, and have their being in eternity, it is impossible for saved fallen souls still in corporeality to imagine themselves *not being*. *For example,* no reader of, or listener to, this book can think back to a time when he or she was *not* (that is, did not exist) or did not have consciousness. This is partly so because all souls existed as ideas in the Supraconsciousness of the Creator-God even before they were created. (If readers or listeners try to think of a time when they did not exist, they will not be able to even imagine it.)

There was an instant in eternity when each one of us was *pushed* into being from idea status to a personal state of volitional self-awareness; it was then that we were *vocalized, articulated,* and *actualized* individually, collectively, and corporately — all at once. And, once created, souls are not able to conceive of a time when they did not exist. It is that simple. As soon as we were brought forth into being, we were *joined* in eternity to eternity. And, regardless of whether we are in an immortal or mortal state of being, our souls remain eternal.

Once souls were created by God, they could not become *uncreated.* In other words, all souls will continue throughout eternity without ever stopping because all souls were created to be eternal. Just as the reader or listener cannot take back their sincere kiss of friendship from a friend who becomes unfaithful to their friendship, so also the Creator-God cannot undo His gift of granting eternity to each volitional and self-aware created being. Although "cannot" might seem like hyperbole concerning the omnipotent Creator-God, part of His gift of eternal life to newly-created beings was His decision that He *would not* ever take the gift back. He imposed that constraint on Himself before He created all souls. This is what makes the gift of eternal life such a remarkable gift. We might destroy the gift given to us personally (whence comes the notion of true freedom with responsibility), but the Creator-God still will not take it back.

That Christ Jesus, the spoken Word, creative Logos, or divine Principle of the Creator-God identifies himself as the "Alpha and Omega" *(Revelation 1:8, 1:11, 21:6, and 22:13)* does not mean that he has a beginning and an ending. Christ Jesus identifies himself as the Alpha and Omega because he is the *First Cause* of the physically-observable universe as well as its *Final Cause:* The physically-observable universe has its beginning and ending in Christ Jesus — which is to say, its creation and re-creation have their origin and completion in him. Christ Jesus is the be-all and the end-all of everything. Christ Jesus is every bit of *it.* (Christ Jesus is not the Theory of Everything, he is the Evidential of Everything.) Christ Jesus is not just *the Way-shower;* he is *the Way.* The entire physically-knowable universe has its restoration to the spiritually-observable universe at the end of all relative space-time. This occurs when God the Father infuses the *all* that belongs to Christ Jesus with His *All,* which is the Totality of His Being. This restoration coincides with the end of the physically-knowable universe.

The Creator-God evolves spiritual ideas in the spiritually-observable universe that are reflected in the physically-observable universe, including all that is ordered and non-random. Cosmic evolution, biological evolution, and consciousness evolution in the physically-observable universe reflect the ideas, thoughts, concepts, and constructs found first in the spiritually-observable universe.

The First, Prime, and Primary Cause of everything (except for Evil, iniquity, and sin) is the Creator-God. He alone is responsible for *the Whole Universe* and its component parts. He alone is responsible for the spiritually- or metaphysically-observable universe; and He alone is responsible for the physically-observable universe. Any and all order in the physically-observable universe is a manifestation of the Creator-God's divine Principle, or creative Logos. Cosmic evolution, biological evolution, and consciousness evolution in the physically-observable universe are not self-induced evolutions or evolutions by random chance or coincidence: not one of them is godless. Except for Evil, iniquity, and sin, the Creator-God is responsible for creating everything. And, except for Satan and his fallen angels, demonic forces, and unclean spirits, the Creator-God is responsible for creating everyone.

The Christian metaphysics of Mary Baker Eddy did not recognize: (1) that there would be a new creation (that is, a *re-creation*) at the end of *the Millennium;* and (2) that there is a necessity for re-creation after the eradication of the effects of all Evil, iniquity, and sin from the entire physically-knowable universe. The Christian metaphysics of Eddy only recognized the restoration of *the Whole Universe* to spiritual sense through spiritual *unfoldment.* For the sake of clarity here, *unfoldment* is different from evolution. *Unfoldment* is the gradual understanding of the truths in the spiritually- or metaphysically-observable universe and their practical applications to the human experience. Unfortunately, *unfoldment* does not include the restoration of fallen, mortal souls to immortality because Eddy's Christian metaphysics does not clearly acknowledge that the truth of all being is found solely in the shed blood of Christ Jesus.

Fallen, mortal beings can only have a finite sense of the eternal. Because of this, they easily misconclude that infinity is the same as eternity. To them, infinity is the same as eternity because infinity "goes on forever." They do not understand that "forever" is a concept that only relates to the space-time of the entire physically-knowable universe, which possesses dimensionality. In contrast, eternity is dimensionless — it is without relative space and without relative time.

In eternity, time *is* space and space *is* time. Further, time and space are absolute only in eternity. They are not uniquely absolute in the physically-observable universe. Here, they are relative. For the sake of clarity, time in eternity is *absolute time,* time in the physically-observable universe is *relative time,* and time in the void beyond the fringes of the physically-observable universe is *empty time.*

In the spiritually-observable universe, *here,* or *absolute space,* is the counterpart to *relative space* in the physically-observable universe; and *now,* or *absolute time,* is the counterpart to *relative time* in the physically-observable universe. To be sure, everyone in the spiritually-observable universe is *here* and *now.* Thus, in the spiritually-observable universe, (1) *here* and *now,* (2) the *here-now,* or (3) *absolute time* and *absolute space* are the counterparts to *relative space* and *relative time* (i.e., *relative space-time*) in the physically-observable universe. Although the Creator-God fills every *place* in the spiritually-observable universe *here* and *now,* the Creator-God does not fill all relative space and relative time in the physically-observable universe. The Creator-God does not reside in the physically-knowable universe. The Creator-God resides only: (1) in the spiritually- or metaphysically-observable universe; and (2) within saved fallen souls still residing in corporeality.

Order and non-randomness in the physically-observable universe are not *perverted* images and *perverted* reflections of the spiritually-observable universe but, rather, *inverted* images and *inverted* reflections. Cosmic evolution and biological evolution are the inverted images and inverted reflections of the Creator-God's expansion of the spiritually- or metaphysically-observable universe. Inverted images and inverted reflections of the spiritually-observable universe are never perverted unless the observer himself/herself has been perverted by Evil, iniquity, and sin. Indeed, inverted images and inverted reflections of spiritual realities and truths are made sense of by the human brain that has been inspired by Christian metaphysics — which is to say, educated, trained, and nurtured in thinking as Christ Jesus thinks.

The uninspired human brain cannot distinguish spiritual realities, or truths, from inverted images and inverted reflections, but the inspired human brain can (the *inspired human brain* is the *spiritually nurtured brain*). Inverted images and inverted reflections of the spiritually-observable universe are interpretable by the inspired human brain. Inverted images and inverted reflections of spiritual objects and truths are made sense of by the inspired human brain as capably as the uninspired human brain makes sense out of the inverted images and inverted reflections of physical objects that fall upon the retina.

It is not the Creator-God's responsibility to create order out of the chaotic ideas to which human beings are exposed. It is the individual responsibility of human beings to try and make sense of it all. *For example,* it is our responsibility to make sense out of creationism and evolution, two substantive yet seemingly-contradictory perspectives. Indeed, we need to pray to our Creator-God for insight, understanding, and wisdom, but the Creator-God wants us to exercise our own free will in making intellectual decisions that help explain seemingly contradictory and opposing ideas, thoughts, concepts, and constructs. Rather than pointing our finger at people with ideas, thoughts, concepts, and constructs that differ from our own, we need to struggle to understand them and, then, (1) accept and integrate them into our own belief systems, (2) mentally shelve them for future consideration, or (3) reject them after carefully considering them.

The following paragraph is a good first starting point for melding creationism and evolution:

Regardless of whether you "believe in" (which is to say, "accept") the paradigm of evolution, its major strength is found in the unifying concept that it presents to the human mind for understanding the interrelationship of all life forms on Earth (and throughout the physically-observable universe). Similarly, regardless of whether you "believe in" (which is to say, "accept") the paradigm of Biblical creationism, its major strength is found in the unifying concept that it presents to the human mind for understanding the basic sequence in the origin of all life forms on Earth. Expressed in these ways, because they are not pitted against each other, we are freed to consider how evolution and creationism can best be interrelated in a unifying paradigm through Christian metaphysics.

There is one Creator-God but tandem creations, or two created elaborations: the spiritually-observable universe and the physically-observable universe. They are parallel universes superimposed on one another because they are on different planes of consciousness. If one lives in the spiritually-observable universe, then that universe is superimposed on the physically-observable universe. And if one lives in the physically-observable universe, then that universe is superimposed on the spiritually-observable universe.

Cosmic evolution, biological evolution, and consciousness evolution are neither false nor unreal; in fact, they represent Creator-driven order and non-randomness in the physically-observable universe. Although each creation (that is, each universe) is no less God-driven than the other, the blind forces of physical nature are quelled in us individually, collectively, and corporately by apprehending the spiritual, metaphysical, and supernatural forces of Spirit.

To summarize at this juncture, order and non-randomness in the physically-observable universe are not perverted images and perverted reflections of the spiritually-observable universe but, rather, inverted images and inverted reflections of the spiritually-observable universe.

>>>>><<<<<

A Note on the Permanent Dissolution of All Corporeality

Regardless of the specific paradigm used for its demise, the physically-observable universe will eventually come to an end. The physically-observable universe will undergo its final phase change when God the Father infuses it with the Totality of His Being after the millennial reign of Christ Jesus on Earth *(1 Corinthians 15:24-28)*. Depending on the paradigm used, either a collapsed and imploded physically-observable universe will be engulfed and expunged by God the Father or a rapidly-accelerating and continually-expanding physically-observable universe will be subdued and dissolved — that is, overtaken and erased — by Him. Independent of the mechanisms involved, the net effect will be the same: one day, approximately one millennium from now, the physically-observable universe will disappear and be replaced with something more closely resembling the Creator-God's original, unfallen creation — something not corporeal but incorporeal in nature. And, although there will be "no more sea" *(Revelation 21:1 KJV)* in this re-creation, the Creator-God's "water of life" *(Revelation 21:6 KJV)* will be present instead. Thus, although the hallmark of all biological life is physical water, the hallmark of all spiritual life is the essence of the Creator-God, which is His Holy Spirit. To be sure, the "pure river of water" in the new creation *(Revelation 22:1 KJV)* is the Creator-God's very own Spirit!

1.3 Thermodynamics

The Laws of Thermodynamics relate to the physically-observable universe but not to the spiritually- or metaphysically-observable universe. Following is a discussion of the three laws of thermodynamics as they relate to the entire physically-knowable universe as well as to the theme of *intelligent evolution:*

1.3.1 The First Law

The First Law of Thermodynamics is also known as the Law of Conservation of Energy. It states that the overall internal energy of an isolated system remains constant.

Energy is simply changed from one form to another in an isolated system. This is the essence of Einstein's $E = mc^2$ where E is energy, m is mass (mass is the amount of matter that an object contains), and c is the velocity of light. Generally speaking, if energy is lost or gained in an isolated system, then there must be a corresponding change in the system's surroundings. However, because the entire physically-knowable universe is infinite, it is an isolated system that has no surroundings; therefore, energy is neither lost nor gained in the entire physically-knowable universe.

In contrast, unlike the entire physically-knowable universe, the physically-observable universe is finite and an open system; therefore, the physically-observable universe can lose energy and matter to its surroundings — which is to say, it can lose energy and matter to the empty vacuum of space that is beyond its fringes (i.e., the cosmic horizon) and constitutes its *surroundings*. For the sake of clarification, astronomers' measurements of redshift and brightness in light emitted at various points throughout the universe show that the physically-observable universe continues to expand into the empty vacuum of space beyond its fringes at a rapidly-accelerating rate. (One might look at this as the physically-observable universe having no outer boundary or an outer boundary that is ever-increasing.) (See a brief discussion of *redshift* on page II-76.)

In a way, the surroundings of the physically-observable universe include not only the vacuum of space beyond its fringes but also the vacuum of space between all ordinary matter that exists within it. Although many physicists would claim that *dark matter* exists in the contiguous vacuum space between matter in the physically-observable universe, this so-called substance could also subsume ghosted images of measurable energy from various subatomic particles, quanta, and electromagnetic radiation that have already passed through the vacuum spaces in between the various clusters of ordinary matter in the physically-observable universe.

It is important to note at this juncture that when physicists claim that *dark matter* is nothing and something at the same time, they are playing a game of semantics because something can never really be nothing and nothing can never really be something — no matter how hard anyone tries to make it so. Nothing is nothing, and something is something, and the two never meet except when the Creator-God creates something out of nothing or makes something out of next-to-nothing. Indeed, the Creator-God creates *ex nihilo* as well as makes *de novo*. (Nuanced differences between the two phrases *ex nihilo* and *de novo* are addressed in Section 3.2.5 in Volume Two of this book.)

Metaphysically speaking, chaos can be regarded as *nothing* because the *something* of chaos is non-ordered and random; thus, the matter and energy of chaos would only be regarded as *something* metaphysically if the matter and energy of chaos gained order and became non-random. Although some physicists might chide creationists for inventing intelligent energy that has ordered the physically-observable universe, the same physicists think nothing, so to speak, of hypothesizing *string theory* to explain the physically unexplainable.

Most evolutionists would reject the notion of the Creator-God's intervention to change nothing into something — which is to say, to change the void and formlessness of the matter and energy of chaos into the order and non-randomness through which cosmic evolution, biological evolution, and consciousness evolution could take place. Instead, they attribute all evolution to random chance and coincidence. And most creationists would reject the notion of an ever-expanding physical universe that originated from *the Big Bang* 13.8 billion years ago because they think that such a notion conflicts with the Genesis account of creation.

Both groups of people have been unable to harmonize evolutionary theory with what they think is the Genesis account of creation. The Creator-God did intercede to bring order and non-randomness to the physically-observable universe during and after *the Big Bang*. In fact, the Creator-God not only interceded 13.8 billion years ago but continues to intercede today. The Creator-God intercedes through the Creator-God's spoken Word — which is His creative Logos, or divine Principle of Creation, Christ Jesus himself. It is the spoken Word of God, Christ Jesus, that creates order out of nothingness and chaos:

> {1} In the beginning, God created the heaven and the earth.
> {2} And the earth was without form, and void; and darkness
> was upon the face of the deep. And the Spirit of God moved
> upon the face of the waters. {3} And God said, "Let there be
> light," and there was light. {4} And God saw the light, that it

was good: and God divided the light from the darkness. {5} And God called the light Day, and the darkness He called Night. And the evening and the morning were the first day.

Genesis 1:1-5 KJV

Here is the present author's rendering of Genesis 1:1-5 that blends creationism and cosmic evolution:

{1} After *the Big Bang,* the Creator-God brought order out of chaos in the physical universe by separating matter and energy from the empty vacuum of space and organizing them: {2} At first, matter and energy had no form and the entire universe had the appearance of darkness because there was no source of physical light-energy in the entire universe. But the Spirit of the Creator-God acted to change the physical appearance of the entire universe by moving His creative Logos, or divine Principle, upon it. This imposed order in the physically-observable universe. {3} Then the Creator-God said, "Let there be physical light-energy," and there was physical light-energy. {4} And God saw that the physical light-energy was good: and the Creator-God divided the physical light-energy from the empty vacuum of space. {5} And the Creator-God called the light-energy "Day," and He called the darkness of the empty vacuum of space "Night." The process of separating matter and physical light-energy from the empty vacuum of space is responsible for the emergence of relative space and relative time in the physically-observable universe. These events constitute the first cosmic eon of relative space-time.

1.3.2 The Second Law

The Second Law of Thermodynamics states that the entropy of an isolated system almost always increases. (This Second Law of Thermodynamics is also known as the Law of Increasing Entropy.) Entropy represents the gradual loss of usable energy in an isolated system, which lost energy results in an increase in disorder and randomness regardless of how uniform the disorder and randomness may become. Chaos (that is, disorder and randomness) in an isolated system almost always results at the expense of order. Order deteriorates, resulting in chaos, or non-order. Therefore, the entire universe (i.e., the physically-knowable universe), which is an isolated system, is headed in the direction of increasing chaos. In order for entropy to decrease in an isolated system, external usable energy needs to be employed to increase and maintain its order. Thus, any order maintained in our solar system in particular, or in the physically-observable universe in general, would come from its surroundings. Since the spiritually- or metaphysically-observable universe constitutes the "surroundings" of the entire physically-knowable universe, any order initiated and maintained anywhere in the physically-observable universe, including our own solar system, originates somewhere in *the Whole Universe* outside of the entire physically-knowable universe. In other words, the effects of all negentropy in the physically-observable universe can be traced to the Creator-God's *theions* described previously in this book. This includes any cosmic and biological order as well as all restoration to order (such as physical healing by spiritual means as well as all other beneficial supernatural events).

1.3.3 The Third Law

The Third Law of Thermodynamics states that the entropy of a system approaches a constant value as the temperature approaches absolute zero. The entropy of an isolated system at absolute zero

(0° Kelvin) is zero. (For reference, absolute zero is defined as minus 273.15^{O} Celsius or minus 459.67^{O} Fahrenheit.)

Since there is no entropy in the spiritually- or metaphysically-observable universe, matter and physical light-energy do not exist there. Theoretically, if matter and physical light-energy could exist in the spiritually- or metaphysically-observable universe, they would be instantly annihilated and disappear as if they had never existed. Just as "flesh and blood cannot inherit the Kingdom of God" *(1 Corinthians 15:50 KJV)* so also can matter and physical light-energy not exist in the spiritually- or metaphysically-observable universe. The Creator-God's *theions* act as a meta-physical kind of antimatter to matter.

Relative to the third law of thermodynamics, at absolute zero no entropy exists in the physically-observable universe, but no entropy ever exists in the spiritually- or metaphysically-observable universe because divine light and divine love self-propagate and, thus, divine energy never dissipates there. This is an important point because it is God-driven negentropy from the spiritually-observable universe that imparts all order and non-randomness to the physically-observable universe. And it is God-driven negentropy from the spiritually- or metaphysically-observable universe that holds everything together in *the Whole Universe* through the very essence of God, which essence is His Spirit — whose function, or root, is the divine Mind, or Supraconsciousness, of God.

To summarize at this juncture, the three laws of thermodynamics are in operation only in the physically-observable universe and are never in operation in the spiritually- or metaphysically-observable universe.

Note: Entropy will again be discussed in Volume Two of this book.

1.4 Genesis Days and Geologic Time

Unfortunately, many people who believe in the authority of the Holy Bible do not see the necessity for harmonizing the Genesis account of creation with reasonable perspectives widely held to be true in the natural sciences. It is unfortunate because seeming inconsistencies between the Genesis account and prevailing views in anthropology, archeology, astronomy, biology, chemistry, cosmology, geology, paleontology, and physics exist mainly because of the ways in which people have been taught to hold the views provided by Genesis and the natural sciences not only at variance but also as irreconcilable. Many Christians are taught to fear evidence from the natural sciences and to believe in a pseudo-science that attempts to validate their doctrinal perspectives and misguided religious conclusions, especially with regard to the timeline of creation. It is equally unfortunate that most natural scientists are taught to seek only natural explanations for all supernatural phenomena reported in the Holy Bible. It is unfortunate because many modern scientists ignore the possibility that some faith-based explanations may not only be valid but are also the only explanations possible.

Because the Biblical account of creation indicates that the Sun, moon, and stars were not created until the fourth *day,*[8] solar time, lunar time, and sidereal (or stellar) time did not exist to measure time for the first three so-called *days* of creation. The eisegesis[9] of conservative theologians would argue that the Hebrew word for *day* (that is, *yom*) always means a twenty-four hour period of time throughout the entire Holy Bible. They fail to take into consideration that there are two verses in the Holy Bible stating that "one day with the Lord is as a thousand years and a thousand

---

8 Genesis 1:14-19

9 *Eisegesis* is defined here as "interpretation with personal, or subjective, bias."

years as one day."[10] They also fail to take into consideration that the planet Earth during its formation had days that were much shorter than they are now. The rotation of the planet Earth about its axis has slowed down considerably since the Earth's formation, and it continues to slow down. The planet Earth's earliest days, consisting of daytime and nighttime, were closer to six hours in duration.

Rigidly narrow Christian theologians would argue that the use of the words "evening" and "morning" for the first three *days* — recorded in Genesis 1:5, 1:8, and 1:13 — reinforces the notion that the *days* in the Genesis account were exactly as they are now. However, without solar, lunar, and stellar light during the first three *days* of creation, there could be no evening, or *setting* of the Sun, and no morning, or *rising* of the Sun. Therefore, either "evening" and "morning" are referring to different referents *(for example,* "evening" could be referring to the beginning of one cosmic eon and "morning" could be referring to its ending) or they are included simply for the purpose of literary parallelism for each of the recorded seven *days* of creation — similar to the factually inaccurate parallelism found in the repetition of fourteen generations three times in the genealogy of Christ Jesus recorded in Chapter One of the Gospel of Matthew. (That some generations have been omitted in Chapter One of Matthew is acknowledged by many Christian theologians.)

Concerning "evenings" and "mornings" for the planet Earth, we should also be reminded that currently there are places on the planet Earth that have no evenings and mornings during any given twenty-four hour period of time because they are situated near one of the two poles. *For example,* throughout much of Antarctica, where the South Pole is located, there were no nights from September 22, 2015 through March 21, 2016 because it was only sunny during that period of time. Thus, in much of Antarctica,

---

10 **Psalm 90:4 & 2 Peter 3:8: King James Version**

because "sunlight" is six months long and "darkness" is six months long, there is only one evening each year and only one morning each year. As a result, if defined as consisting of one evening and one morning, one "day" in cities near the South Pole is one year long and not twenty-four hours long.

Certainly, it is not impossible to reconcile the seven *days* of the Genesis account of creation with prevailing views in the natural sciences if one recognizes the validity of three concepts: (a) that each of the seven Genesis "days" represents a substantially longer period of time — what the present author calls a *cosmic eon;* (b) that there is a "fast forward" presentation of creation events in Genesis; and (c) that the Creator-God can slow down or speed up time at His Will:

1. Many conservative Christians are frightened by the concept that the Genesis account of the seven days of creation may not consist of twenty-four hour days because they erroneously believe that such thinking might take away from the believability of the Biblical message of salvation through Christ Jesus alone. They unconsciously or consciously subscribe to the notion that everything in the Bible must be true literally, or exactly as they understand it, or nothing in the Bible can be true. Paradoxically, as a result, their own faith in the entire Biblical narrative is found wanting.

2. A "fast forward" perspective is a concept understood by many conservative Christians. *For example,* in explaining the seventieth week of Daniel *(Daniel 9:24-27),* some conservative Bible students and scholars skip time and resume counting when relevant events begin again in the future. I make this point not to endorse their views on the seventieth week but to indicate that the "fast forward" concept is embraced by people who are not labeled as heretics by other conservative Christians. Another example of a "fast forward" is found in between verses 23

and 24 in 1 Corinthians 15, where the Apostle Paul "fast forwards" 1,000 years from the time of Christ Jesus' return (verse 23) to the time that he delivers his kingdom to God the Father (verse 24).

3. That God can slow down or speed up time is a concept accepted by many conservative Christian theologians — *for example,* when they acknowledge that God slowed down time to honor a prayer request from King Hezekiah of Judah *(2 Kings 20:8-11)* or that God stopped the sun and the moon from moving to honor a command from Joshua *(Joshua 10:12-13)*.

As the present author sees it, the inability of some people to reconcile seemingly contradictory and/or complex details, concepts, and facts in the Genesis account of creation with prevailing views in the natural sciences concerning evolution is due to their failure to think metaphysically, or conceptually.

Christian metaphysics is a tool that can be used much like binoculars, enabling students of life to carefully study details of both accounts from afar (i.e., objectively) that can then be blended together to form a coherent narrative and unified theory of creation-evolution. The act of harmonizing and blending creationism and the theory of evolution is our responsibility and not the Creator-God's responsibility. Because we have the basic facts, it is our responsibility to put them together into something that is intelligible and honors both perspectives. *For example,* a metaphysical harmony could be achieved between the Genesis account of the creation of Adam and prevailing views of evolution in the natural sciences if the Genesis account of the creation of Adam represents the original creation of immortal beings in an incorporeal Paradise known as the Garden of Eden at the same time that a race of hominins without souls was evolving at the Creator-God's direction on the planet Earth. Such harmonization would posit that when errant spiritual beings fell to temptation and were concomitantly expelled (i.e., exiled) from their immortal state and incorporeal condition, their souls became "frozen" in a state of being (which state is *mortality*) that included the relative space-time of the physically-observable universe. In other words, their appearance in corporeality, or human flesh, was coincident with, and dependent on, their fall and expulsion from their original, glorious estate in God's Paradise (i.e., the Garden of Eden). This harmonization would help to explain how Cain and Seth, the second and third sons of Adam and Eve, were able to find wives who were not their own siblings. Their spouses would have come from the race of hominins without souls who had evolved biologically at the Creator-God's direction. (The presence of intelligible speech or intelligible sign language is an indicator, but not the only indicator, of the presence of a soul in a corporeal body.)

Understanding this harmonization also requires the capacity to conceptualize that the unfallen Adam was actually a composite of spiritual beings — the majority of whom had to wait until after their collective fall for their individual turn to enter corporeality through having their souls housed temporarily in preassigned human bodies. ("Temporarily" here refers to the actual lifespan of individual human beings.)

If you, the reader or listener, are *only* looking to find flaws in the metaphysical harmonization just presented, then you are missing the major point. Regardless of whether the harmonization just presented by the present author is precisely accurate or not, harmonization should be attempted in order to show that these two bodies of knowledge (one body of knowledge that is supernatural and the other body of knowledge that is natural) can be complementary and not opposing or contradictory.

When one uses metaphysics as a tool to harmonize the schemata of two different conceptual frameworks, one should not expect there to be an exact one-to-one correlation and perfect alignment between comparable or contrasting sets of elements from the two frameworks. In fact, skillfully and methodically using Christian metaphysics reveals that, when taken together, various schemata from the two frameworks may not only be true at the same time but also can be superimposed over one another to reveal a greater truth. To be sure, simultaneously attending to multiple layers of truth produces a vision of the whole that is significantly greater than the sum of its parts. Practically speaking, using Christian metaphysics enables one to understand truths that are supernaturally overlaid. The present author can attest that the tempo of one's understanding gradually quickens from *adagio* to *allegro* as one routinely employs Christian metaphysics to look at life.

Based on the genealogies carefully recorded throughout the Old Testament, or *Tanakh* (the Hebrew Bible), it is clear that Adam and Eve were people who lived *approximately* 6,000 years ago. Of course, the genealogies in the Old Testament are tedious reading but necessary in order for us to calculate the *approximate* passage of time since the appearance of Adam and Eve in corporeality — which is to say, in human flesh.

Thus, whatever else the Holy Bible is or isn't, it is a book that covers 7,000 years of time on Earth (that is, seven *days* of 1,000 years each):

• 4,000 years from the appearance of Adam and Eve in corporeal flesh at the time of their "fall" (expulsion and exile from Heaven) to the first advent of Christ Jesus (the passing of four "days");

• 2,000 years from the time of the first advent of Christ Jesus to the second advent of Christ Jesus (the passing of two "days");

• 1,000 years for the millennial reign of Christ Jesus on Earth (the passing of one "day"), culminating in: (1) World War IV (known as the Battle of Gog and Magog in the Book of Revelation), (2) the Great White Throne Judgment of the

Creator-God, and (3) the creation of "a new heaven and a new earth" *(Revelation 20:8, 20:11, and 21:1)*.

The majority of Christians and modern scientists should be able to agree on the chronology presented in the Holy Bible relative to a 7,000 year period of Earth time. If that can be a second starting point for mutual understanding and agreement between Christians and natural scientists, then a majority of the desired harmonization will have been achieved (see page I-40 in this book for the first starting point).

It is clear from the Holy Bible that, in addition to *Adam* being the name of a historical person, the name *Adam* is also a plural word representing: (1) humanity as a whole, (2) *Homo sapiens* in general, and (3) an entire specific group of hominins with fallen souls. *Adam* is a Hebrew plural word for male and female corporeal beings with an iron-based, or *reddish,* pigment. (*Reddish* here is referring directly to the hue of iron-containing oxygenated blood and, thus, only indirectly referring to related skin color.)

To summarize at this juncture, the seeming variances between the Genesis account of creation and the facts and well-grounded theories in the natural sciences concerning cosmic evolution, biological evolution, and consciousness evolution can be reconciled harmoniously:

1. *If* Christians and natural scientists would read Genesis 1:1 through Genesis 2:7 as a condensation, or capsulization, of astronomical, geological, chemical, biochemical, and biological events that include the eventual emergence of an entire species of hominins without souls whose physical bodies were capable in complexity of housing the fallen souls of the original Adamic race of spiritual beings.

2. *If* Christians and natural scientists would read Genesis 2:8 through Genesis 2:25 as a description of the creation

of a heavenly, or incorporeal, paradise known as the Garden of Eden and the creation of incorporeal beings collectively referred to as *Adam*. This premise presupposes that the Garden of Eden was (and still is) in a parallel, incorporeal world superimposed over and above the planet Earth in a different plane of consciousness — specifically in eternity and not in the space-time of the physically-observable universe.

3. *If* Christians and natural scientists would read Genesis 3:1 through Genesis 3:24 as a condensation, or capsulization, of the fall to temptation of individually-created, incorporeal beings — collectively known as *Adam* — that resulted in their expulsion from an immortal state of being to a mortal state of being.

Although physical coordinates are given for the Garden of Eden in Genesis 2:10 through Genesis 2:14, those coordinates can be understood metaphysically as *also* representing a locus in a parallel, incorporeal world superimposed over and above the planet Earth in a different plane of consciousness. That is why Christ Jesus referred to the Kingdom of God as "at hand" *(Mark 1:15 KJV and Luke 21:31 KJV)* — meaning, "within us," "right next to us," and "beside us." This is also why the heavenly Paradise of God is described as a garden with trees in Revelation 2:7, 22:2, and 22:14. Indeed, as mentioned previously, the Creator-God's Garden of Eden and His heavenly Paradise are synonymous.

Metaphysically speaking, are Christian people not *trans-species?* Here, I am *not* writing about dysphoric people or contemporary pagan people sometimes referred to as *furries* (i.e., people who fancy that they possess the spirits of animals other than *Homo sapiens*). I am writing about fallen, albeit saved, created beings who feel like aliens on the planet Earth because they know their true home is in a different state of being.

(1) Are saved human beings not spiritual beings living in corporeal bodies? (2) Are we not strangers in a strange land? (3) Will we not leave our human bodies behind one day? (4) Will saved fallen created beings in the future not have new bodies, or refreshed somatic identities, that will be more representative of who they really are in Christ Jesus? If you can answer "Yes" to the four questions just posed, and if you can picture, imagine, and understand what the four questions represent, then you are thinking about the future using Christian metaphysics.

If you can think metaphysically about the future that is described in Revelation 21:1 to 22:5 of the Holy Bible, then you should be able to think metaphysically about the ancient past that is described in Genesis 1:1 to 3:24 of the Holy Bible as well.

Part Two
Bridging the Gap between Creationism and Evolution: Using the Tool of Metaphysics as a Problem-Solver

2.1 Thinking Metaphysically

The question "What is truth?" has been debated for millennia. The truth be told (and it is being told right here), there are different levels of truth, but the highest level of truth for human beings is Christian metaphysical truth. Although Christian metaphysics is both spiritual and supernatural, it is insufficient to only define Christian metaphysical truth as *spiritual truth* or *supernatural truth.* The following paragraphs in this section help to define *Christian metaphysics* more fully by discussing how it is involved in seeing, thinking, knowing, and believing.

People in the physically-observable universe need to see metaphysically — that is, they need to be able to discern the cause, substance, essence, meaning, and purpose of physical phenomena — including physical objects, events, and conditions of animate being. It is in this way that they see past the physically-observable universe to the spiritually-observable universe. In contrast, people in the spiritually-observable universe do not need to work toward seeing metaphysically because they automatically and clearly see the cause, substance, essence, meaning, and purpose of everything with which they come into contact mentally. In other words, people in the spiritually-observable universe do not need any special tool, not even the tool of metaphysics, to observe their own reality nor, for that matter, the reality of mortal beings. For immortals in the spiritually-observable universe, knowledge itself provides inner sight. For the saved souls still inhabiting the physically-observable universe, their authentic faith provides spiritual sight, including metaphysical hindsight, insight, and foresight. However, for lost souls in corporeality (i.e., unsaved human beings), only their physical senses provide sight — but only an outer sight that permits them to experience physical reality and not the eternal reality to which immortal beings belong. Here, let us be reminded that no human eye has seen nor human ear heard what the Creator-God has prepared for those who love Him and wait upon Him *(1 Corinthians 2:9 and Isaiah 64:4).*

What human beings see as the physically-observable universe is an altered version of the spiritually- or metaphysically-observable universe. It is as if human beings are looking at the spiritually-observable universe through a kaleidoscope whose viewing chamber has been rotated so that all images from the spiritually-observable universe are twisted, bent out of shape, and refracted at a disadvantageous viewing angle. Thus, the reality that human beings see is different from the reality that actually exists within the spiritually-observable universe because the viewing chamber is fashioned from *iniquity,* which is a result of our collective turning from obeying the Will of the Creator-God. It is as if we are looking through a metaphysical black hole where iniquity serves as the gravitational field in its tunnel that distorts all images from the spiritually-observable universe. Seen in this way, the physically-observable universe constitutes the "shadow of turning" *(James 1:17 KJV)* — *turning* here synonymous with "iniquity," and *sin* defined within the present author's literary works as "action based on that turning."

It is only through *re-turning* to the Creator-God by consciously accepting the shed blood of Christ Jesus as the only atonement for our iniquity and sin that human beings can catch glimpses of the spiritually- or metaphysically-observable universe while their souls are still held in corporeality. Personal suffering and living in a perpetual state of contrition can increase the definition of images from the spiritually-observable universe by increasing the resolution of what is metaphysically seen.

When *meta* and *physics* were first linked together relative to the works of Aristotle *(Aristoteles),* they did not have the same combined meaning that the word *metaphysics* has today. Before his book entitled *Metaphysics* was published, Aristotle wrote a series of eight "books" referred to by the opus title **Φυσικὴ ἀκρόασις** *(phusike akroasis)* — literally, "nature orations" or "lectures [about] nature" — and gradually referred to simply as *Physics* (i.e., *Nature*). Aristotle's written work entitled *Physics* (or *The Physics*) is a collection of writings on *natural philosophy* *(natural philosophy* is the noun phrase that predates *natural science)* with an emphasis on many topics that have as much to do with metaphysics as they do with the modern science of physics. In fact, Aristotle's *Physics* greatly influenced Pierre Teilhard de Chardin's elaboration of the final teleological cause and purpose that he named the *Omega Point.* (See Section 2.5.4.2 — entitled *The Psychism of de Chardin* in Volume Two of *Intelligent Evolution* for more on the *Omega Point.*)

After *Physics,* Aristotle wrote a series of fourteen "books" that were referred to by the opus title **τὰ μετὰ τὰ φυσικά** *(ta meta ta phusika)* — literally, "that [which was written] after *the Physics"* and gradually referred to simply as *Metaphysics* (or *The Metaphysics).* The word *metaphysics* eventually took on the connotation of "that which is beyond the physical" — or "that which is invisible, including that which is spiritual or supernatural" — and has increasingly grown in acceptance as such.

Although the word *metaphysics* is not in the active vocabularies of most Christians living before *the Millennium,* the present author has tried to lay the groundwork for its greater acceptance and use after Christ Jesus returns. Here are a few of the definitions for *metaphysics* that the present author has given in some of his previous works:

> *Metaphysics* here means "a spiritual science and sense beyond comprehension by mere human science and sense." *(As I See It: The Nature of Reality by God, page 107, Footnote 218)*
>
> *Metaphysics* describes the nature of reality. *Christian metaphysics* describes the nature of spiritual reality from the standpoint of salvation through Christ Jesus. *Metaphysics* resolves things into thoughts and thoughts into things. *Christian metaphysics* accomplishes the same thing except that every view is filtered through the lens of the Holy Bible with all hindsight, insight, and foresight provided by the

only teacher of all truth, the Creator-God's Holy Spirit. Thinking metaphysically for Christians requires that we hold the whole spiritual truth while simultaneously attending to its various parts. Thinking metaphysically for Christians also requires that we look beyond corporeality and physical explanations to spirituality and supernatural explanations for understanding how to resolve life's challenges. *(God, Our Universal Self: A Primer for Future Christian Metaphysics, page xi)*

Christian metaphysics for the third millennium of the Christian era is a way of looking at life that recognizes and acknowledges the existence of a supernatural reality and a spiritual universe in addition to the existence of a corporeal reality and a physical universe. However, contemporary Christian metaphysics employs the understanding that a supernatural reality and its accompanying spiritual universe supersede any and all physical, material, or corporeal realities without denigrating the practicality of physical, material, or corporeal solutions to physical, material, or corporeal problems. *(God, Our Universal Self: A Primer for Future Christian Metaphysics, page 88)*

To be sure, in order to use principles of Christian metaphysics to treat life's problems, we must first confess that the Lord Jesus Christ is in control of everything. *(God, Our Universal Self: A Primer for Future Christian Metaphysics, page 94)*

For the purpose of this book, *metaphysics* is defined as "the nature of reality," consisting of *ontology* (i.e., the study of being and existence), *natural theology* (i.e., the study of God and how God relates to this world and the things in this world), and *universal science* (i.e., the study of ultimate principles and how they impact our understanding of causality and our understanding of the levels of organization of matter and their interactions as well as their finitude). By extension, *divine metaphysics* [i.e., *Christian metaphysics*] is therefore defined as "the *nature* of supernatural reality" (i.e., the *essence* of spiritual reality). To be sure, divine metaphysics [i.e., *Christian* metaphysics] has no physical bounds except in its description. *(Divine Metaphysics of Human Anatomy, pages 8-9)*

Divine metaphysics [i.e., *Christian metaphysics*] (noun phrase): (a) the nature, or essence, of spiritual, or supernatural, reality; (b) that which is beyond explanation based on natural science or the laws of physics, chemistry, and biology; (c) that which resolves things into thoughts, concepts, ideas, and principles as well as that which resolves thoughts, concepts, ideas, and principles into things [i.e., spiritual objects] based on spiritual and supernatural reality and sight (i.e., spiritual insight, hindsight, and foresight). *(Divine Metaphysics of Human Anatomy, page 22)*

Metaphysics is the study of unseen realities. *Metaphysics* is also the nature, or essence, of the highest spiritual reality. *Metaphysics* includes the understanding that thoughts are things and things are thoughts. Metaphysics takes into account that there is a spiritual universe in addition to a physical universe. *Spiritually-scientific metaphysics* does not negate that there is a physical universe. Instead, it takes into consideration that there is a higher reality of which an understanding is necessary in order to effect reproducible spiritual changes in various human conditions. Metaphysics employs spiritual truth to effect emotional, mental, physical, spiritual, and social change. *Christian metaphysics* is the highest form of metaphysics. *(Hello from 3050 AD!, page 87)*

As you stand firm in your understanding of the power and authority of the shed blood of Christ Jesus, you are employing Christian metaphysics. *Standing firm* in that shed blood is beyond being literal or figurative: it is metaphysical. It is metaphysical because it is based on faith in informed

> ways. ("Standing firm in the shed blood of Christ Jesus" would be figurative only if it were used in a poetic sense by someone who is merely offering lip service to its truth or blithely commenting about it.) *(Hello from 3050 AD!, page 94)*

Anytime that you think beyond the physically-observable universe and beyond physical activities and *conditions of being* associated with them, you are thinking metaphysically. Thinking metaphysically includes thinking conceptually about what is hidden to one's physical senses. Thinking metaphysically from a Christian standpoint requires searching for a higher, greater, and invisible reality that can be experienced only through the heightened and elevated spiritual sense that is derived supernaturally from the Creator-God's Holy Spirit that indwells us.

If you draw a graph on paper using x, y, and z axes and can imagine how the graph would look three-dimensionally in infinity and you assign meaning to the graph, then you are thinking metaphysically. If physical objects represent concepts to you *(for example,* if an upholstered armchair represents *relaxation* to you), then you are thinking metaphysically. If certain concepts are tangible to you because you can clearly imagine them, then you are thinking metaphysically. If you recognize that someone who is hurling insults at you or speaking sarcastically to you is trying to stab you and cut you to the emotional and spiritual quick, then you are thinking metaphysically. If you can think of interrelated concepts as intersecting geometric shapes *(for example,* as correlated factors represented in a Venn diagram), then you can think metaphysically. Thinking metaphysically is thinking outside of the box where the box was only an idea to begin with. Although quantification may not occur when you think metaphysically, qualification always does. In other words, you may not think in terms of numbers and percentages using quantities, but you will always think in terms of characteristics and descriptive aspects using qualities.

If you believe that behind each physical thing and every physical experience there is at least one associated invisible concept, then your belief system is grounded in metaphysics.

If you believe that you are alive in Christ Jesus, then you are thinking metaphysically. If you look for spiritual reasons or causes to explain the situations and circumstances you are in, then you are thinking metaphysically. If you say "I see" when you finally understand a difficult concept, then you are expressing yourself metaphysically.

In order to indicate their frame of reference clearly, people who think metaphysically may try to qualify the source of their spoken and written thoughts with phrases like: "Metaphysically speaking," "Spiritually speaking," "Supernaturally speaking," "Humanly speaking," "Physically speaking," "Corporeally speaking," "From a metaphysical standpoint," "From a spiritual standpoint," "From a supernatural standpoint," "From a human standpoint," "From a fleshly standpoint," "From an earthly standpoint," "From a physical standpoint," "From a corporeal standpoint," "From a physically-natural standpoint," and "From a spiritually-natural standpoint."

If you believe that you are whole and healthy in Christ Jesus regardless of an unchanging or worsening physical condition, then you are applying metaphysical principles to your daily life. If you are a Christian who happens to be paraplegic and you claim that you walk daily with Christ Jesus and that you are running a race to please the Creator-God, then you are looking beyond appearances to a higher reality that is metaphysical, and not physical, in nature (here, *in nature* means "in essence").

One of the reasons to study Christian metaphysics is to help us separate legitimate spiritual thinking from thinking steeped in dogma, superstitions, mythologies, urban legends, and folk tales. Thinking about the omniscience, omnipotence, and omnipresence of the invisible Creator-God always involves metaphysical thinking.

Yes, the Creator-God came to Earth in the form of Christ Jesus in order to: (1) help us understand eternal truths; (2) present God to us in a more relatable way; and (3) experience firsthand what it is like to be mortal and human. Additionally, the Creator-God came to us in the flesh as the *only-begotten Son of God* to teach us the difference between: (1) what has real value and what has no value at all; (2) what pleases God the Father and what displeases Him; and (3) how to behave and how not to behave. Christ Jesus is our eternal role model and our eternal mentor through the Creator-God's Holy Spirit. Of course, first and foremost, *God the Son* gave his life for us as the only substitutionary offering acceptable to *God the Father* for our iniquity and sin. (Christ Jesus is not just the *Way-shower,* Christ Jesus is the *Way.*) Salvation through the *only-begotten Son of God* can only be understood fully by thinking and conceptualizing metaphysically.

Many people do not grasp the meaning of *begotten* in the expression *only-begotten Son of God.* Therefore, for the sake of clarity, it is important to state here that "begotten" is the past participle of the verb "beget," whose past tense is "begat" *(beget, begat, begotten).* The word *beget* means "to give birth to" or "to bear" *(bear, bore, born).* Thus, the word "begotten" means "born," "birthed," or "physically conceived by the union of reproductive cells (i.e., spermatozoon and oocyte) and delivered at parturition from a uterus." The first man Adam was not "begotten" by the God of the Holy Bible. Only Christ Jesus was "begotten." In the case of Christ Jesus, "begat by God" and "begotten by God" mean that: (1) God Himself provided the seed and Mary (Miriam) herself provided the egg for Christ Jesus to be conceived; and (2) Christ Jesus was physically born through Mary's birth canal, consisting of uterus and vagina. Christ Jesus was generated: (1) by God the Father not through sexual relations but through His Holy Spirit *overshadowing* Mary *(Luke 1:35 KJV);* and (2) by Mary the mother through her personal physical contributions of egg, uterus, and intrauterine nutrition. Although Mary is the mother of Jesus, and Jesus is God-in-flesh (God Incarnate), Mary is not the mother of God.

Thinking metaphysically goes way beyond thinking in figurative or poetic language or only using one's imagination. For Christians, thinking metaphysically requires: (1) an imagination that is tethered to the Creator-God through His Holy Spirit; (2) looking for and applying general spiritual truths and principles to daily living; and (3) expressing our thoughts in language that seeks to keep our own consciousness elevated at the same time that it seeks to elevate the consciousness of others without policing and correcting the ways in which others express themselves. (However, it is okay to police and correct one's own thinking and how one expresses oneself.)

Thinking metaphysically includes recognizing that life was incorporeal before it became corporeal and that we were created as spiritual beings first. Thinking metaphysically enables us to catch a glimpse of the absolute truth here and there, reminding ourselves at the same time that, in corporeality, we can only see and know in part. It helps us to realize that, although the Creator-God is everywhere, He is not to be found in physical objects. It causes us to think of the Sun, moon, planets, stars, solar systems, galaxies, and the entire physically-observable universe as representations of spiritual concepts, principles, and ideas in the mind of God. Metaphysical thinking even permits us to conceive of parallel universes — one spiritual and the other physical — existing side by side — each superimposed on the other.

If you see a butterfly and can imagine that it represents a flying flower in God's spiritually-observable universe, then you can think metaphysically. If you can look at water as a physical representation of the Creator-God's Holy Spirit, then you can think metaphysically. If you can conceive that human corporeal images, appearances, or forms mask our compound, composite, collective, and corporate identities in Christ Jesus, then you can think metaphysically.

When you learn a spiritual principle, you are actually learning it metaphysically; and, if you try to apply it practically to relevant situations and conditions, then you are seeking to apply it metaphysically. When you think metaphysically, you understand that you are on a spiritual journey and that you are either moving toward or away from the Creator-God. Thinking metaphysically for human beings requires us to use earthly tools such as alphabet letters, individual words, phrases, and sentences, but we should also be reminded that written language first originated as pictographs and ideograms. Thus, modern communication requires us to think, speak, and write using contemporary alphabetic language that metaphysically represents the inter-relationship of pictures, ideas, concepts, constructs, and images.

Thinking metaphysically includes the recognition that spiritual principles build upon one another and that, once we learn one major spiritual principle, we are then better prepared to learn the next one. Thinking metaphysically requires refinement throughout one's life by living in spiritual inquiry combined with one's unending gratitude to the Creator-God for everything that we have and all that we are. Thinking metaphysically enables us to select important concepts from the belief systems of others and accommodate and assimilate them into our own belief systems. Thinking metaphysically and expressing ourselves metaphysically permit us to hand down important thoughts, ideas, concepts, and constructs from one generation of learners to the next.

If you regularly look for an invisible reality behind the physical appearances that you see, then you are thinking metaphysically. If you acknowledge that there is a hidden, invisible reality behind the motives of others, then you are thinking metaphysically. If you understand that Evil often masquerades itself as Good and that you may be fooled by Evil, then you are thinking metaphysically. If you ask God to refine your ability to discern elements of His supernatural reality, then you are seeking to understand life metaphysically.

When you think metaphysically, you gradually become more aware of the cold, dead images that come from Satan's mortal mind and the crisp, animated images that come from the Creator-God's immortal Mind. You cannot think of the Lord God Almighty without thinking metaphysically, but thinking metaphysically requires spiritual nurture and daily practice in thought, in mind, in word, and in deed.

Although the word *metaphysics* may not be in the active vocabularies of most pre-millennial Christians, *metaphysics* as a process is regularly used by most pre-millennial Christians. *For example,* following are two statements that are not only believed and understood by pre-millennial Christians but also believed and understood metaphysically — even though they may not know or understand the specific word *metaphysics:*

#1 Christ Jesus bore the iniquity and sin of the world on the cross of his crucifixion at Calvary.

#2 The shed blood of Christ Jesus is the only sacrifice acceptable to God the Father for the remission of our sins and for His forgiveness and removal of our iniquity.

In statement #1, it is physically true that Christ Jesus was crucified on the cross at Calvary. However, although every authentic Christian understands and believes that it is literally true that Christ Jesus bore the iniquity and sin of the world on the cross at Calvary, they also understand that it is not physically true that he "bore their iniquity and sin" because: (1) iniquity and sin are not physical objects; (2) iniquity and sin have no physical mass; and, therefore, (3) iniquity and sin cannot be physically transferred or carried (i.e., borne). And, since it is not figuratively, or metaphorically, true that Christ Jesus bore the iniquity and sin of the world, then it can only be metaphysically true — that is, true spiritually as well as supernaturally. Thus, in order for authentic Christians to understand statement #1, they must be using metaphysical thinking even though they may not know or understand the specific word *metaphysics*.

In statement #2, it is physically true that Christ Jesus shed blood when he was crucified on the cross at Calvary. However, although every authentic Christian understands and believes that it is literally true that the shed blood of Christ Jesus remits our sins when we accept him as Savior and grants us forgiveness and removal of our iniquity by God the Father, they also understand that it is not physically true that his blood remits our sins or grants us forgiveness and removal of iniquity because: (1) "sin" is not a ledgered item in an accounting column; (2) iniquity is not a physical object; and, therefore, (3) forgiveness of sin and removal of iniquity are not physical actions. And, since it is not figuratively, or metaphorically, true that the shed blood of Christ Jesus atones for our sins and grants us forgiveness for our sins and removal of our iniquity, then such atonement and forgiveness can only be metaphysically true — that is, true spiritually as well as supernaturally. Thus, in order for authentic Christians to understand statement #2, they must be using metaphysical thinking even though they may not know or understand the specific word *metaphysics.*

If authentic Christians are already using metaphysics without knowing or understanding the word *metaphysics,* then why should it be important for them to be taught what the word means?

Knowing and understanding the word *metaphysics* is important for the following reasons:

1. Unless students of life know and understand *metaphysics* as a branch of philosophy and theology that is worthy of study, they will not be properly educated, trained, and nurtured in its practical applications nor be able to help others become properly educated, trained, and nurtured in its practical applications.

2. Unless students of life know and understand *metaphysics,* they will not be able to systematically learn its basic principles — which principles can then be

employed to help resolve earthborn challenges and solve earthborn problems like unemployment, underemployment, spousal difficulties, poor health, disabilities, preparing for the future, survivability, sustainability, thrivability, and financial challenges.

3. Unless students of life know and understand *metaphysics,* they will not be able to properly evaluate any systematic theology that claims to incorporate principles of metaphysics.

4. Unless students of life know and understand *metaphysics,* they will not be able to actively and proactively use its principles to answer questions associated with truth, reality, causality, purpose, and being nor use its principles to effect emotional, mental, spiritual, physical, or social change in their lives.

To be sure, metaphysics and its principles could be studied using new nomenclature, but, then, we would be neglecting the important legacies of those who have spent quality time and effort in studying and articulating their views concerning metaphysics not only as a branch of philosophy but also as a branch of theology.

To summarize at this juncture, it is important to emphasize that Christian metaphysics involves thinking conceptually with Christ Jesus at the helm of one's thoughts, taking and holding "captivity captive" *(Ephesians 4:8 KJV).* In short, Christian metaphysics elevates one's thinking.

2.2 What Thinking Metaphysically Is Not

Thinking metaphysically does not mean that you become so heaven-bound that you are no earthly good. It does not mean that you are so lost in thought that you shirk your daily responsibilities and duties. It does not mean that you use figurative language to impress others or to puff up the image you have of yourself. It does not mean that you play "word police" or "thought police" in order to fix, or correct, the colloquial speech and individual thinking of others so that they do not appear to express negativity or support what you think is incorrect or in error.

Thinking and expressing oneself metaphysically is not creating jargon or using slang to trivialize another person's difficult situation or circumstance — *for example,* (1) using the word *invalid* as an adjective ("not valid") to signify the seeming unimportance of a debilitating condition in an *invalid* (noun that means "disabled person"); or (2) using the hyphenated word *dis-ease* to refute an authentic disease process, denigrates standard medical protocols used to treat it, and trivializes the severity of the condition's consequences. To be sure, thinking and expressing oneself metaphysically is not creating jargon to diminish a fundamental of the Christian faith — *for example,* using the hyphenated word *at-one-ment* with God without elucidating that unity with God comes only through the atonement of Christ Jesus. The guideline for jargon and slang that the present author uses and recommends to other metaphysicians is that if a slang or coined word or phrase takes away significant, intended meaning from a word or phrase and cannot retain its uniqueness when translated into a different language, then the slang or coined word or phrase should not be used or used only with a thorough explanation.

In contrast to the previous two paragraphs, thinking metaphysically allows the thinker to understand that everyone is on his or her own personal journey and that we each can share the spiritual concepts and ideas to which we subscribe in a calm and courteous manner, looking: (1) to retain the concepts and ideas of others that resonate within our souls and are complementary to our understanding of Biblical principles; and (2) to discard the concepts and ideas of others that are harmful or not helpful to us and/or make light of, or denigrate, the Creator-God's absolute truth.

Thinking metaphysically to solve problems and resolve issues is not mind control. It is not disciplining one's mind to control external situations and circumstances. It is disciplining one's mind to control internal reactions and responses to life's situations and circumstances. It is not the mind of one human being controlling the mind of another human being. It is each mind yielding willingly to the Will of the Creator-God in order to please Him at the same time that each mind desires and seeks to know the thoughts, ideas, and thinking of the Creator-God on specific topics, subjects, and issues.

Thinking metaphysically is not fantasizing how we might like something to be or what we might like someone to become.

Thinking metaphysically to solve problems and resolve issues is not the "Word of Faith" movement, the "Speaking Things into Existence" movement, the "Confessing It and Possessing It" movement, or the "Prosperity" movement (all four religious movements are essentially the same).

For those who are reading or listening to this book during *the Millennium,* and who may not be familiar with the nomenclature just used, the "Word of Faith" movement was a huge pyramid scheme where those at the top of the pyramid (church leaders) struck it rich and those at the bottom of the pyramid (congregational members) were left playing the Creator-God as if He were a programmed gambling device (i.e., slot machine) that always pays out when the right formula is employed. The congregants were told by their leadership that if they were not prosperous, then they were using the wrong Biblical formula or thinking negatively. This near-perfect deception blamed poverty and poor health on the negative thinking and negative vocalizations of those who were not materially prosperous or not in good physical health. Victims of physical disability, disease, senescence, poverty, unemployment, underemployment, abuse, and murder were blamed for their own predicaments. Victims were taught to be ashamed of their situations and circumstances. Their conditions were labeled a source of embarrassment for themselves, their leaders, and their fellow congregants.

The phrases "speaking things into existence" and "confessing it and possessing it" became quite trendy and cultish; they were in vogue during the latter part of the twentieth century and the early part of the twenty-first century. The phrases were included in a religious movement that cut and pasted different Bible verses together to build a false doctrine. It put people in bondage so that they could more easily be manipulated by their church leaders. Proponents of this movement often used verbiage from the following Bible verse: "It is God who calls those things that do not exist as though they are" *(Romans 4:17b KJV Paraphrase)*. Customarily, these proponents omitted the portion that references the Creator-God, who had created the physically-observable universe and all living things in it by speaking them into existence *(Genesis 1:1-27)*. Advocates of "speaking things into existence" and "confessing it and possessing it" taught that all Christians should be able to call or confess into existence advantageous situations that do not currently exist because such calling and confessing is the Will of God.

In effect, the "Prosperity" movement teaches that:

1. Poor people do not have enough faith, are negative in their thinking, and fail to employ prosperity-related Biblical principles.

2. Unhealthy people do not have enough faith, are negative in their thinking, and fail to employ healing-related Biblical principles.
3. Disabled people do not have enough faith, are negative in their thinking, and fail to employ ability-related Biblical principles.
4. Persecuted people do not have enough faith, are negative in their thinking, and fail to employ freedom-related Biblical principles.
5. Murdered people did not have enough faith and were guilty of (a) not declaring that only good would come to them as well as (b) not listening to the voice of God warning them about impending harm.
6. Christians are entitled to get what they want because they deserve it and because God wants them to have it.
7. The material prosperity of human beings is more important to God than their humility.
8. To manipulate and exploit God successfully, one needs to know the right formulas, think positively, and employ Biblical principles.
9. Church members must submit to the authority of church leaders because they are God's chosen representatives on Earth and because they can teach them the right Biblical formulas to use in order to obtain what they want.

The worst thing about the "Word of Faith" movement is that it ends up blaming victims for their own victimization. This perversion of God's written Word is represented in the following thinking:

> "If only they had listened to God, they would not have been in that predicament."
>
> "They must have expressed negativity in their thinking."

"They should have spoken protection into existence."

"They must have received what they deserved."

"They got what they were asking for."

"They brought it upon themselves."

"They reaped what they sowed."

"They lacked faith in God."

The sins of the "Prosperity" Movement include:

1. Taking the Creator-God for granted.
2. Failing to pray "Thy Will be done."
3. Lacking gratitude for what one already has regardless of whether it is a little or a lot.
4. Placing material prosperity, worldly success, and physical healing above salvation in importance.
5. Not taking the Creator-God at His full Word (using only some Bible verses and not others).
6. Teaching others false doctrine (less than whole, or less than balanced, doctrine).
7. Rejecting what others have to say if it does not follow a prescribed formula or ritual using word-specific, cultish verbiage.

In summary, the entire "Prosperity" movement is a perversion of "laying claim to what is rightfully ours." Unfortunately, Christians often confuse their fleshly mind (the mind of mortal man) with the mind that they have in Christ Jesus. Rather than trust the Creator-God to know what is rightfully theirs, they trust themselves to decide instead. They fail to recognize that "the mind governed by the flesh is hostile to God; it does not submit to God's law, nor can it do so" *(Romans 8:7 KJV)*.

Unfortunately, the representative thinking of the "Word of Faith" movement is not much different from the judgmental thinking expressed by the "friends" of Job. The "Word of Faith" movement is not that far afield from snake handling, incanting spells, magical thinking, and stage performing. It discourages people from diligently studying the whole Bible, thinking for themselves, and exercising personal free will — all three of which are activities that please the Creator-God because they utilize the gifts that He has given to us. In the final analysis, the "Word of Faith" movement reflects the original sin of Adam and Eve, who fell to Satan's temptation for them to "be as gods" *(Genesis 3:5 KJV)*.

2.3 A Cautionary Note

In their efforts to identify and explain *the unseen unknown — for example:* intelligence, consciousness, God, divine Mind, Christ Consciousness, and/or the Supraconsciousness of God (not all of which are synonymous here nor in most other places) — some spiritually-minded people may misinterpret the reasons for: (1) the navigation instincts of certain insects, fish, birds, reptiles, and mammals *(for example,* honeybees, sockeye salmon, homing pigeons, loggerhead sea turtles, and bottlenose dolphins); (2) the sexual attraction of various animals to their potential mates that are a few kilometers away *(for example,* corn earworm moths); and (3) the detection of minute quantities of blood by certain predators *(for example,* great white sharks). In their efforts to identify *the unseen unknown* and attribute the examples cited to *mystery,* some spiritually-minded people may unwittingly ignore the roles of sensing magnetic fields (i.e., magnetoreception) as well as olfactory reception of diffusing pheromones and other chemicals in gaseous and aqueous media. In other words, they may ignore the role of organismic taxes (i.e., behavioral responses to external stimuli [pronounced tak-seez′]) in the responses of animate life forms to their changing physical environments.

The present author's scientific research for his Master's thesis investigated the nuclear polyploid composition of *Physarum polycephalum,* a common slime mold. Although some people with a metaphysical bent might attribute this mold's *pushing, probing,* and *groping* for a nutrient supply to its tapping into a soul force intelligence or life force consciousness, this organism is simply reorienting itself and altering its form (in an alternation of generations) based on its changing physical environment.

The overall lesson here is that, although there can be a metaphysical interpretation for the origin, existence, actions, and interactions of biological life, Christian metaphysics should never be used to explain unusual biological phenomena as part and parcel of occultic forces when there are ways to understand these phenomena physically. Metaphysically-minded people should avoid creating or indulging a mystery religion from their own ignorance and pseudoscience or from the ignorance and pseudoscience of others.

2.4 Proposed Curriculum for *the Millennium*

As the present author sees it, the curriculum for secondary and tertiary school students during *the Millennium* should include the following sequenced coursework related to metaphysics:

1. The History of Metaphysics
2. The Theology and Metaphysics of Aristotle and Plato
3. The Theology and Metaphysics of Immanuel Kant

4. The Theology and Metaphysics of Georg Hegel
5. The Theology and Metaphysics of Mary Baker Eddy
6. The Theology and Metaphysics of Pierre Teilhard de Chardin
7. The Theology and Metaphysics of Joseph Adam Pearson
8. The Theology and Metaphysics of Christ Jesus

Offered in tandem with the sequenced coursework related to metaphysics should be the following sequenced coursework in the mathematical and natural sciences:

1. The History of Natural Philosophy/Natural Science
2. Aristotelian Logic, the Philosophy of Mathematics, and Computer Programming (Applied Mathematics)
3. Algebra/Geometry (Euclidian Geometry)
4. Geometry (non-Euclidian Geometry)/Trigonometry
5. Advanced Algebra/Calculus
6. Physics/Astronomy/Cosmology
7. Chemistry/Geology
8. Biology/Anthropology/Paleontology

2.5 Insights, Implications, & Applications from Others

Why has the present author looked to Aristotle, Kant, Eddy, de Chardin, and Hawking for insights, implications, and applications for his book on *intelligent evolution?*

A newborn baby is not conscious of being a baby. It may be aware of other things around it, but it is not self-aware. Self-awareness is not initiated until an infant begins to learn and understand:

(1) language labels for other people in its presence, (2) language labels for physical objects to which it is exposed, (3) language labels for experiences that it has, and (4) language labels for potential experiences that it may have. An infant only begins to become conscious of itself in an intelligent way when it begins to perceive that other people, objects, and experiences are not its *self* (i.e., its identity) and that it is not other people, objects, or experiences. To be sure, before it has language labels for people, objects, and experiences, a baby is aware that it is hungry or satiated, tired or awake, thirsty or slaked, warm or cold, comfortable or uncomfortable, etc., but it does not yet have a rudimentary understanding of who, what, where, and when it *is* — which is to say, it does not comprehend its own *being*. As an infant develops cognitively, beginning to understand the meaning behind language labels constitutes the awakening of its consciousness concerning its own human condition and being. Such an understanding represents the baby's becoming aware of itself as a living being in contrast to other people, things, and experiences.

Understanding language labels, then, provides the key to unlocking a baby's awareness of itself. Understanding language is the tool to unlock such self-awareness for the baby as much as understanding the metaphysical meanings behind language provides the tool necessary to unlock our comprehension of: (1) the physically-observable universe; (2) the spiritually-observable universe; (3) *intelligent evolution* — including cosmic evolution, biological evolution, and consciousness evolution; and (4) the Supraconsciousness, or divine Mind, of the Creator-God. In other words, just as a baby's beginning to understand the meaning of language labels helps it to become aware of its conscious functioning self in contrast to the world around it, so does our beginning to understand the metaphysical meanings behind language representing people, objects, and experiences help us to become aware of our own supraself, or higher self, which is the absolute identity that we have in our Creator-God through Christ Jesus. It is in grasping metaphysical meanings that we are granted an acute, or refined, understanding of the hidden meanings behind our own lives and individual, collective, and corporate *being*.

The right language, the right language labels, the right syntax, and the right semantics mean everything to our individually understanding the principles of *being* and the meaning of life in relationship to the Supreme Being. That is why education, training, and nurture are so important. Human beings who do not receive the highest levels of education, training, and nurture will not fulfill their full potential for understanding themselves or their Creator-God. A spiritually enlightened sense derives metaphysical meaning from language describing objects, actions, and experiences; unfortunately, a developed intellectual and cognitive sense without a spiritually enlightened sense derives only physical meaning from language describing objects, actions, and experiences.

In our education, training, and nurture, we look to the language of others to see if it can provide us with insights, implications, and applications concerning our own *being* and the reality, or realities, in which we find ourselves. That is why the present author has looked to the works of Aristotle, Kant, Eddy, de Chardin, and Hawking. These authors were not only superior thinkers who thought metaphysically, they also had superior skills in articulating their views. To be sure, each of the authors cited did not have a perfect grasp of all metaphysical meanings of life and *being,* but their thinking and writing help to provide us with ideas, concepts, constructs, and language labels applicable to the subject of *intelligent evolution*.

Thinking metaphysically causes desired intellectual outcomes to materialize and sought-after ideas, concepts, and constructs to crystallize within one's own understanding. Thinking metaphysically about *intelligent evolution* requires deep thinking. And deep thinking involves: (1) reflecting on the topic to comprehend what is known about it as well as what is not known about it; (2) juxtaposing the topic with other topics to align, overlie, gird, and/or brace them with one another; and (3) opening oneself to ideas about the topic from others.

If alive today, (1) Aristotle would probably identify himself as a philosopher and a naturalist but not a religionist, (2) Kant would probably identify himself as a philosopher but not a naturalist or theologian, (3) Eddy would probably identify herself as a metaphysician and a theologian, (4) de Chardin would probably identify himself as a philosopher-theologian as well as a natural scientist, and (5) Hawking would probably identify himself as a theoretical physicist, cosmologist, and atheist.

Let us now turn to the contributions of Aristotle, Kant, Eddy, de Chardin, and Hawking to clarify our own thinking relative to *intelligent evolution*. To be sure, the literary efforts of these five thinkers require and inspire deep thinking through the insights they provide as well as the various implications and applications they stimulate — regardless if any of these authors intended their literary efforts to engender thinking beyond their own or not.

In the following sections, the present author will be integrating his own thinking with the thinking of Aristotle, Kant, Eddy, de Chardin, and Hawking.

2.5.1 Insights, Implications, and Applications from Aristotle

Reading or listening to Aristotle's collections of writings entitled *Physics* and *Metaphysics* is like panning for gold. It is tedious work, but occasionally you find a speck of truth that makes your efforts worthwhile, especially as it relates to a historically early metaphysical understanding of chance, change, temporality, eternity, corporeality, and incorporeality.

As indicated previously, the two book titles, *Physics* and *Metaphysics,* are primarily *transliterations* of Greek words and only secondarily *translations* of those same words. The titles might be more accurate in English if they were, respectively, *About Nature* and *Beyond Nature.* And, although these collections of Aristotle's writings may be referred to as *books,* it probably would be better to call each of them a compilation of *orations,* or *discourses,* because Aristotle's explications were intended to be read as lectures, or at least used as notes for lectures, to audiences that included students and colleagues in his Peripatetic school at the Lyceum of Athens, Greece. With this said, the works could also be titled *Discourses About Nature I* and *Discourses About Nature II.*

None of the written material in Aristotle's *Physics* and *Metaphysics* perfectly reflects modern science or the contemporary understanding of the scientific method employed by modern scientists and represented in the following eight steps:

1. Formulating a question.

2. Performing a background investigation.

3. Constructing an original hypothesis.

4. Testing the hypothesis through experimentation.

5. Analyzing results from the experimentation.

6. Drawing a conclusion by accepting or rejecting the original hypothesis.

7. Communicating the results and their implications.

8. Formulating subsequent questions.

At one point, Aristotle wrote of an *approach* to discovery that ended with a *conclusion,* but at no point is that approach detailed in formulaic steps. Because of the lack of certain methodologies at the time, it would be more accurate to refer to Aristotle as a *natural philosopher* and to Aristotle's *natural philosophy* (that is, Aristotle's *philosophy on nature*) rather than use the modern verbiage of *natural scientist* and *natural science* to describe Aristotle and his work. Outside of some basic algebra, geometry, and physics that Aristotle used to provide proofs for a few of his hypotheses, Aristotle relied heavily on rational argumentation, including deductive and inductive reasoning, and the "evidence of our senses" *(Physics, VIII.8, p. 217)* to demonstrate intellectual experimentation for his hypotheses and conclusions. At times, Aristotle's hypotheses were null hypotheses, accepted or rejected on the basis of the results from thought-experiments using refutation, argument, and critical analysis without the benefits of modern statistical tools. To be sure, Aristotle's methodology is only a precursor of the modern scientific method.

Although his first collection of writings is called *Physics* and his second collection is called *Metaphysics,* evidence of Aristotle's thinking on physics and metaphysics is found throughout both works. To be sure, Aristotle did not use the word *metaphysics,* and he himself did not provide the true etymological basis for the meaning of that word. Instead, what we consider *metaphysics* Aristotle referred to as *First Philosophy, wisdom, theology,* and *the science of being as being* (the last of which might be reworded today as *the science of being itself,* or *the science of being in itself* — which is to say, the reality of *being* alone and not in relation to any qualifiers). Today, *the science of being* is simply referred to as *ontology.*

Aristotle's writings in *Physics* and *Metaphysics* are tedious to read, especially during one's first reading of them. Indeed, one might call Aristotle's writings *elliptical* and another might call them *rambling.* Periodically, the truth is somewhere in between. To be sure, the reader of Aristotle's works must be attentive in order to catch glimpses of metaphysical truth that Aristotle's writings offer. Although Aristotle's writing style might remind some modern readers of Gertrude Stein's literary style, Aristotle's style is more pedantic and less staccato than Stein's and, of course, each of these thinkers (Aristotle and Stein) had a different intent for the repetition of their slightly amended phrases and sentences. (Stein's successive alterations in repetitive sentence structure really had their origin in her capacity to look at life kinetoscopically.[11]) Additionally, just as poems each have their own cadence, so do the prose styles of most writers have their own cadence. At times, Aristotle's written cadence reminded the present author of auction chanting, especially when Aristotle tried to prove his hypotheses through highly redundant rational argumentation.

Philosophy in general and metaphysics in particular play important roles in helping us to resolve seeming theological conflicts. Although formulating a thesis, its antithesis, and their synthesis is attributed to Johann Fichte (1762-1814), one can find the underpinnings to such an approach in resolving conflict by paying attention to Aristotle's use of two *contraries* and their *intermediary,* which approach is included in both his *Physics* and *Metaphysics.*

Although Aristotle lived before Christ Jesus and the origin of Christian writings that are now collected together to form the 27 books of the New Testament, using Aristotle's philosophical approach can easily resolve such Christian doctrinal conflicts as: (1) the tri-unity and oneness of the Creator-God; (2) the two different spoken formulas used for water baptism; (3) the initial moment of the indwelling of the Holy Spirit in the new Christian convert; (4) geocentrism and heliocentrism; and (5) the interpretation of the Genesis account of creation in relation to

---

11 The Kinetoscope, patented by Thomas Alva Edison in 1897, was one of the earliest motion picture devices that permitted one viewer at a time to peer through a small hole and watch a succession of still images merge into what appeared to be movement. The Kinetoscope was a prototype of the cinematic projection system.

cosmic, biological, and consciousness evolution. Aristotle would have described any biases held for each of the five conflicts just named as the *affections* of an individual believer's mind. He would have presented the opposing views as clearly as possible and then looked for a way to articulate an intermediary view that would have appeased those on either side of each issue.

To be sure, without various philosophical and metaphysical approaches to describe why Christ Jesus is the *only-begotten* Son of the Creator-God, and why his crucifixion holds eternal significance for all people, no person who intends to avoid reading the Holy Bible (because that person is contentious, illiterate, or both) could ever be converted to Christianity. Philosophy in general and metaphysics in particular can be helpful to explain difficult theological concepts, especially to people who are not comfortable with the language of the Holy Bible, regardless of its specific translation or version. Yes, we should allow the Holy Bible to speak for itself, except to those who are unable to read it or are unwilling to listen to it being read aloud.

Aristotle's *Physics* and *Metaphysics* are appropriate for both a secondary school curriculum and a tertiary school curriculum, especially if excerpts from them are read aloud by students — in part, because the writings were intended to be used in orations. The present author would especially recommend Book V from *Physics* and Chapter 7 in Book Lambda from *Metaphysics*. Book V from *Physics* is an important intersection of mathematics, philosophy, and nature. And Chapter 7 in Book Lambda from *Metaphysics* provides equivalency between the *Prime Mover* — also referred to by Aristotle as the *Primary Mover, First Mover,* and *First Principle* — and the Creator-God (that is, the Supreme Being, or Deity).

The following two subsections on Aristotle are devoted to insights, implications, and applications from Aristotelian thinking in *Physics* and *Metaphysics* that especially have relevance to the present author's paradigm of *intelligent evolution*.

2.5.1.1 Aristotle's *The Physics*

For Aristotle, *being* generally refers to perceptible, or sensible, things that are either inanimate or animate. In contrast, for many metaphysicians from Plato *(Platon)* onward, *being* refers to the spiritual idea, spiritual nature, essence, spirit, soul, or *pneuma* of one's inner self, higher self, supraself, or spiritual self in an unseen reality. The closest that Aristotle comes to a contemporary metaphysical definition for *being* is in his *being as being* ("being *qua* being") — which is just another way of referring to "*being* by virtue of itself," or "*being* of, in, and through itself."

For Aristotle, nothing comes into *being* (that is, into material existence) from *non-being*. In other words, all that *is* has come from something else that has preceded it in physical existence. And, except for Aristotle's *Prime Mover,* everything changes due to principles and causes associated with material substance. This has relevance to our modern concept of evolution in the cosmos as well as in life on the planet Earth. To Aristotle, cosmic evolution, biological evolution, and consciousness evolution would be inherent in the very substances of which their physicality is composed. Aristotle would say: "The attribute of changeability is inseparable from what it is an attribute of" *(Physics, I.5, p. 19)*. In other words, if Aristotle had been aware of protoplasm and not just *flesh,* he probably would have concluded that protoplasm also has the capacity to evolve because of its intrinsic nature of changeability.

Although Aristotle viewed his Prime Mover as *Good,* he also viewed it as an impersonal force, not a personal being; and he viewed that something opposite to the Prime Mover — that is, Evil — also exists. In Aristotle's criticism of the Platonists, he states:

> The point is that while *our* view, in the context of there being something divine and good and desirable, is that the opposite to this also exists, as does that which by its own nature desires and longs for it, *they* [the Platonists] are

> committed to the view that the opposite longs for its own destruction.
>
> *Physics, I.9, p. 31*

For the present author, what Aristotle ascribes to the Platonists raises these two important questions: (1) Is it not possible that Satan, the Fallen Lucifer, longed for his own destruction simply by choosing to be the eternal Enemy of the Creator-God? (2) And, although Lucifer was not ever, and is not now, omniscient, did he not possess enough higher order intelligence to know that his rebellion would bring him to utter destruction? At some level, I think the answer to both questions is "Yes." And I think that this conclusion is supported by the verse in Revelation that teaches us that Satan is now hard at work because he knows that his freedom is soon over *(Revelation 12:12 KJV)*.

The major difference between Platonists and Aristotle is that Aristotle was a pantheist (that is, one who believes that the Creator-God can be found in the material universe and not just in its intelligent design). In contrast, Platonists believed that the *Greatest Good* could be found existing independently of matter, regardless of the *Greatest Good's* exact relationship to matter and matter's various appearances. So, whereas Platonists (as well as Neoplatonists) would look for the Prime Mover outside of the material universe, Aristotle and his disciples would expect to find the Prime Mover as part of the material universe itself. However, it should also be shared that thinking between the Aristotelians and Platonists was not always so sharply divided. *For example,* Aristotle believed that the Prime Mover — or anything eternal, for that matter — could not be located in relative time.

Many modern natural scientists are Aristotelian in their approach to the meaning of life even if they might not appreciate or understand the descriptor *Aristotelian.* Why? They would agree with Aristotle that what we can observe is more important than what we are not able to observe. On the other hand, Platonists (as well as Neoplatonists) believe that what we are not able to observe physically is more important than what we can observe physically. Whereas Aristotle and his followers looked for ways to physically measure reality, Plato and his followers looked for ways to gain insights about *universals* as *abstract ideas* and for ways to describe them as *ideal forms* — commonly referred to with an upper case *F* to connote their transcendent nature (i.e., "Forms"). And Aristotle and his followers would look to gain insights about *universals* in *things* (that is, in matter and its substance). As evidenced in their writings, Kant subscribed to Aristotelian realism, Eddy subscribed to Platonic realism, de Chardin was somewhat divided between the two, and Hawking could be described as an Aristotelian and a metaphysical pantheist *(metaphysical* here in an intellectual but not a spiritual sense). (Again, not all Platonic and Aristotelian views are mutually exclusive.)

Describing the branch of philosophy known to him as *First Philosophy* (what is now more commonly referred to as *metaphysics*), Aristotle wrote that "it takes a single branch of knowledge to know the purpose or end of something and the way in which the purpose is achieved" *(Physics, II.2, p. 37)*. In this way, Aristotle captured the essence of teleology. Aristotle added: "From one point of view we too [that is, human beings] are ends." So, if Aristotle knew anything about the biological evolution with which we are now educated (or not educated, as ignorance would have it), he might have concluded that the whole purpose to the physically-observable universe would be to end with the genus and species of *Homo sapiens*. In other words, every other genus and species fulfills what the Creator-God had in mind in order to culminate in and support the emergence, survival, sustainability, and thrivability of humankind. This does not contradict the Holy Bible in that the Genesis account of creation ends with the creation of individual *Homines sapientes* (i.e., human beings) as creation's pinnacle. Because Aristotle would have agreed that the deviser of any plan is a cause, he might well agree with Christians that the Author of the Plan of Salvation through Christ Jesus is the *First and Final Cause* as well as the *First and Final Principle* (or creative *Logos*).

For the sake of clarity, while Aristotle argued that the *telos,* or end, of an acorn is to become an oak tree, the present author would add "for the ultimate purpose of sustaining and enhancing human life." In other words, the *telos* of an acorn is to become an oak tree for the ultimate purpose of providing shelter, food, fire, aesthetics, and ecosystem sustainability for human beings.

Aristotle minimized the roles of chance and spontaneity in the changes that occur in the material universe. He stated that "both chance and spontaneity are... coincidental causes" and that "their sphere of operation is [in] events which do not have to happen" *(Physics, II.5, p. 46)*. As noted by the present author earlier in *Intelligent Evolution,* the physically-observable universe is finite but the physically-knowable universe is infinite. That the empty vacuum of space beyond the fringes of the physically-observable universe is infinite could be classified as coincidental because it is neither a requirement for cosmic evolution, biological evolution, and consciousness evolution nor a requirement for the desired end result of *Homo sapiens* and the interdependent ecosystems contributing to the physical survival of *Homo sapiens.* After explaining chance and spontaneity, Aristotle concluded: "The upshot of this is that however much spontaneity is the cause of the [material] universe, intelligence and nature are bound to be more primary causes..." *(Physics, II.6, p. 48).*

The present author believes that Plato would agree with Aristotle's statement that:

> ...there are two kinds of sources of natural change, and in one kind the source is not itself a natural object, in the sense that it does not contain its own source of change. In this latter category comes anything which causes change without itself changing *(for example,* that which is absolutely unchanging and is the primary entity in the whole universe) and what a thing is, or its form (since that is its end or purpose).
>
> *Physics, II.7, p. 49*

Insights, implications, and applications of Aristotelian thinking include the following: What *Homo sapiens* is, and what physical form it has, is the end or purpose of cosmic evolution, biological evolution, and consciousness evolution. Most of what else exists, including all other biological life, is neither by chance nor by coincidence because most of what else existed, and now exists, fostered the emergence, biological success, and continued survival of *Homo sapiens*. This, of course, presumes that *Homo sapiens* is the only suitable habitation for souls that have fallen from immortality to mortality. To be sure, no other species has the cerebral capacity and capability to channel an eternal soul — neither *Gorilla gorilla* (the Western lowland gorilla) nor *Pongo borneo* (the orangutan) nor *Pan troglodytes* (the chimpanzee) nor *Pan paniscus* (the bonobo) nor *Tursiops truncatus* (the bottlenose dolphin). Plato might add to these insights, implications, and applications that the physical form of *Homo sapiens* in some way reflects the abstract idea, or Form, of Man — capitalized here to distinguish original, or immortal, man from fallen, or mortal, man. (Other than: (1) words that follow certain rules of grammar and syntax, (2) proper nouns in English, and (3) all nouns in German, capitalized words in the fields of philosophy and theology refer either to transcendent ideas or to aspects of Deity.)

To Aristotle, final causes are crucially important in nature because everything has a purpose. He stated: "Now, 'nature' is ambiguous in that it can refer either to matter or to form [i.e., physical appearance]; but since the end is form, and everything else takes place for the sake of the end, it is this form that is the cause since it is that for which everything happens" *(Physics, II.8, p. 52, brackets mine)*. Thus, because *the end is the cause,* we can conclude that *Homo sapiens* is the final physical cause, or physical end, of all evolution. (The final metaphysical cause, or metaphysical end, of *Homo sapiens* is the salvation of souls.) In other words, *Homo sapiens* is the desired result of all evolutionary changes for the purpose of providing opportunities for salvation. However, that *Homo sapiens* is the final cause or end should not be misconstrued to mean that *Homo sapiens* is the intangible Prime Mover — the one true and only real Creator-God. And, just as the form of a brick wall is not its purpose for existing, so too is the form of *Homo sapiens* not its purpose for existing. As explained by *intelligent evolution,* the purpose of *Homo sapiens* is to temporarily house fallen eternal souls that they might be led to repentance. (At physical conception, all eternal souls in corporeality are *unsaved* fallen souls.) Indeed, the form and purpose of *Homo sapiens* are linked, but they are not the same.

To the present author, Aristotle's *change* is the underlying principle for cosmic evolution, biological evolution, and consciousness evolution. Taking Aristotle's statement that "change is the actuality of what is potential" *(Physics, III.1, p. 58),* and applying it to all evolutionary processes, brings us to the conclusion that the cosmos and all of its elements always had the potential to evolve into human life — in keeping with "something which causes change without being changed itself" *(Physics, III.1, p.58)* — or, in other words, in keeping with the direction provided by the Prime Mover, the Creator-God Himself. Indeed, evolutionary change resulting in *Homo sapiens* "is a special kind of actuality" *(Physics, III.2, p.59)*. Concerning the noun phrase *Prime Mover,* it is important to note that the Creator-God can only be referred to as "the Prime Mover" in, and for, the physically-knowable universe. In contrast, the Creator-God can only be referred to as "the All" in, and for, the spiritually- or metaphysically-observable universe.

The present author is in agreement with Aristotle that "the infinite cannot have an origin because that would limit it" *(Physics, III.4, p. 64)* — although the present author prefers the word "eternal" be used instead of "infinite" when speaking of the Creator-God. According to what has already been posited by the present author, it is only the emptiness of space that is infinite in the physically-knowable universe, and the physically-observable universe contained inside of the emptiness of its space is finite. So, *the Big Bang* that occurred from the immense mass of finite centralized energy took place inside the infinite vacuum of space, and, as soon as energy was blown apart from its original central core, the physically-knowable universe came into existence and continued, as it continues today, to expand into an infinite vacuum of space, such space the spot where God is *not.* Paradoxically, nothingness is infinite but not eternal. Here, I must remind the reader that the present author does not use the words *infinite* and *eternal* interchangeably although, at times, Aristotle does. To the present author, *eternal* refers to: (1) that which is outside of the relative space-time continuum of the physically-observable universe; and (2) the *here* and *now* inside of the spiritually- or metaphysically-observable universe. At this juncture, I believe that Aristotle might add: "They also call it the divine [that is, the *Eternal],* on the grounds that it is immortal and imperishable" *(Physics, III.4, p. 64, brackets mine)* — which is in agreement with the Christian theologies of Eddy and de Chardin but not the agnosticism of Kant or the atheism of Hawking.

Aristotle made an important point when he stated that "anything with a source is dissolved back into the source it has come from" *(Physics, III.5, p. 68).* This point is in keeping with: (1) the concept of the eventual permanent dissolution of all corporeality as well as (2) the Creator-God's infusion of the physically-knowable universe by the Totality of His Being at the very end of relative space-time *(1 Corinthians 15:28)* — when Christ Jesus returns the *all* (over which he has been granted power and control) to God the Father, the "All" Who then becomes "All-in-all."

Aristotle was at variance with those who think of the infinite as matter because, as he stated, "it is odd for them to make it [i.e., matter] the container rather than the contained" *(Physics, II.7, p. 76, brackets mine).* In comparison, the present author has already stated that it is *the Whole Universe* that *contains* both the spiritually- or metaphysically-observable universe as well as the physically-knowable universe until the physically-knowable universe is infused by the Totality of the Creator-God's Being and, thereby, ceases to exist.

To reiterate, the present author has identified the empty vacuum of space beyond the fringes of the physically-observable universe as part of the physically-knowable universe. Aristotle identified this empty space as *void.* Concerning this *void,* Aristotle stated that "those who claim that void exists are really talking about place since what they mean by 'void' is probably place deprived of body [that is, place deprived of physical form]" *(Physics, IV.1, p. 79, brackets mine).* Thus, the empty vacuum of space beyond the fringes of the physically-observable universe is really place deprived of matter. That this infinite void is deprived of matter is coincidental and incidental (though not accidental) to the concept of *intelligent evolution:* The infinite void is of no consequence in the long run because it is really nothing at all. Figuratively speaking, it is as if the entire physical creation was painted on a blackboard of empty space.

According to Aristotle, a change of place is known as movement *(Physics, IV.1, p. 78).* To the present author, Heaven, or Paradise, may be an immortal state of being without the place, but it can never be the place without the immortal state of being. For physicality, *place* is anywhere that can be identified by coordinates. Since Heaven, or Paradise, is dimensionless and, therefore, has no coordinates, *place* in the spiritually- or metaphysically-observable universe is, simply, *here* (it is *there* if you are still in corporeality but *here* if you are already restored to eternity). Taken with Aristotle's point that a change of place is known as movement, the following question presents itself: "Is there no movement in Heaven?" The answer is that movement in Heaven is something entirely unlike movement in the physically-observable universe: Because one is always *here* in Heaven, one can never be displaced or replaced there. In other words, in Heaven one simply moves from *here* to *here.* When Ezekiel saw the faces of the cherubim, he was able to see all four of the faces at the same time even though each face was pointing in a different direction from the other three *(Ezekiel 1:10 and 10:14).* The Prophet Ezekiel was actually *seeing through* to the dimensionless *here* and *now* of eternity.

The present author is amazed that so many of Aristotle's statements — when taken in isolation and not in the context of the material, or physical, universe about which Aristotle mostly writes — have great bearing on thinking metaphysically. So, not only are we panning for gold in seeking to find absolute truth, we are also panning for gold in seeking language labels to express absolute truth in contexts different from the contexts in which they were originally used by Aristotle. *For example,* the following statement, though meant to apply to the *place* of a physical object in the material universe, could just as well apply to the so-called *place* of an object in the spiritually- or metaphysically-observable universe:

> ... place will not in fact be a stable entity, and so one place will occupy another place, and there will be a plurality of coincident places.
>
> *Physics, IV.4, p. 87*

Is Heaven not *a plurality of coincident places?* Are the three partitions of the Triune God not coincidental in the *here* and *now* of Heaven? Do immortal beings not translocate from one coincidental place to another as they travel in Heaven? Because Aristotle defines *place* for a physical object in the material universe as "the limit of the containing body" *(Physics, IV.4. p. 87),* could that definition not also be used to define *place* for a spiritual being in the spiritually-observable universe? And, because the containing body of the spiritually-observable universe is limitless, could potential places for a spiritual being not be limitless as well? Are immortal beings in the spiritually-observable universe not in multiple places all at the same time? In eternity, are spiritual beings not always *here* and *now?*

Imagining a moving ship on a river, Aristotle stated that "the container functions as a vessel rather than as a place" *(Physics, IV.4, p. 88)*. Could this not also be said of the somatic identity that we will each have one day in Heaven as restored immortal beings with new bodies? Then and there (in the *here* and *now*), our somatic identities will be the vessels for our souls. In other words, our individual somatic identities will not be *places* for our souls; they will be vessels, or vehicles, for them.

The ideas in the last two paragraphs are metaphysical applications from the natural to the supernatural that are just as sound as taking the Sun in our solar system to represent the Creator-God. (The Sun *is not* the Creator-God; it merely can *represent* the Creator-God — the central core of all life, power, and *being* — whose face, or appearance, human beings cannot look at in proximity without being physically annihilated.)

Ideas, insights, and understanding that we receive from the Holy Spirit give *inner voice* (i.e., mental expression) to the thinking of our Creator-God as well as help to shape our own thinking individually, collectively, and corporately. We communicate with our Creator-God through our thinking, and our Creator-God communicates with us through His thinking. And the individually created beings of God communicate with each other through the sharing of their thinking, regardless of where they are in *the Whole Universe* (i.e., in Heaven or on Earth). True communication between us and God — and among us individually — is an exchange of ideas, insights, and understanding. Metaphysically speaking, our *inner voice* gives shape to the ideas, insights, and understanding we receive and they help to shape our *inner voice.* Moreover, the ideas, insights, and understanding we receive altogether provide the metaphysical form (i.e., Form) for our *inner voice,* and our *inner voice* helps to form, as well as inform, our thinking.

Referring to objects in the material universe, Aristotle stated: "It is also reasonable that everything of its own nature stays in its own place" *(Physics, IV.5, p.90)*. The same could be said when comparing beings in the spiritual universe to beings in the material universe, all of whom stay in their respective domains. Beings in the physically-observable universe do not move to the spiritually-observable universe unless they are *translated* there by the Creator-

God. Neither do beings in the spiritually-observable universe move to the physically-observable universe unless they fall there (i.e., are exiled there) or, like some angels, purposely step into its plane of consciousness.

When Aristotle first discussed time, he emphasized that "the now is not a part of time because a part measures the whole and the whole must consist of its parts; time, however, does not seem to consist of nows" *(Physics, IV.10, p. 103)*. In other words, Aristotle posited that there can be no linear sequences of *nows* in the physically-observable universe or the spiritually-observable universe (using the present author's language labels). He stated clearly: "There is no next now" *(Physics, IV.10, p. 103)*. However, Aristotle later argued against this in *Physics, VI.1-VI.3, pp. 138-146 and VI.6, p. 154.* Regardless of the arguments for or against, in the physically-observable universe the *nows* of the past no longer exist and the *nows* of the future do not yet exist. However, in the spiritually-observable universe, it is always *now.* In other words, there will never be an infinite number of *nows* in the spiritually-observable universe because only one eternal *now* exists there.

The genius of Aristotle is especially demonstrated in the following statement concerning time when he refers to what many today might call a *multiverse.* (To be sure, Hawking could have written this statement.)

> ... if there were a plurality of universes, the movement of any one of them would be time, just as much as the movement of any other one of them, and the upshot would be a plurality of simultaneous times.
>
> *Physics, IV.10, p. 104*

Aristotle's conclusion that time "must be an aspect of change" *(Physics, IV.10, p. 105)* is relevant to the paradigm of *intelligent evolution,* especially in reference to when the major events of *intelligent evolution* took place. Because relative time and relative space are associated only with the physically-observable universe and not with the spiritually-observable universe, evolutionary change through speciation is only associated with the physically-observable universe and not the spiritually-observable universe. Although the spiritually-observable universe is dynamic, its members do not evolve into new species because relative time and relative space do not exist there (only a *here and now* exists there); and its members do not have the potential to evolve into new species because they are already fully actualized. (As stated previously, however, created beings in Heaven — like the Creator-God — always retain the capacity to expand in consciousness.) In contrast, the members of the physically-observable universe have evolved in speciation across relative time and relative space. Biological evolution, however, does not *jump* from one species to the next species; rather, the potentials of biological evolution already exist all at once — as in the wave of the Creator's hand and the utterance of the Creator's voice across relative time and relative space. If the wave of the Creator's hand could be frozen in time and space frame by frame *(for example,* kinetoscopically), we would see new biological species in each successive frame. We might misconclude that the new species came from the prior when, in metaphysical reality, they were all created in succession as parts of one action by the Creator-God — in graded steps across the backdrop of the physically-observable universe's time and space.

All of what has been stated in the previous paragraph is complementary to Aristotle's assessment of where change takes place:

> Evidently, then, anything eternal, in so far as it is eternal, is not in time: it is not contained by time, nor is its existence measured by time. This is indicated by the fact that it is not affected at all by time either, which suggests that it is not in time...
>
> *Physics, IV.12, p. 111*

> So anything which does not change, and does not rest either, is not in time.
>
> *Physics, IV.12, p. 111*

The *here* and *now* of the spiritually- or metaphysically-observable universe is like an ocean, capable of carrying vessels on it and transporting them to regions heretofore not experienced by them. In this case, the transportable vessels include the souls of those in corporeality who are indwelt by the Creator-God's Holy Spirit. Unfortunately, when souls in corporeality are not completely anchored in Christ Jesus, their imaginations are tossed about on such mental journeys. In contrast, when souls in corporeality are completely anchored in the Christ Jesus of "yesterday, today, and forever" *(Hebrews 133:8 KJV),* their imaginations are able to venture forth without fear, anxiety, or trepidation.

It is important for students of Christian metaphysics to use their active imaginations to catch glimpses of higher truths with the single requirement that they be anchored in a Biblical understanding of Christ Jesus. Without that anchoring, their use of active imaginations can lead to disastrous results because, thus untethered, the spirits of such souls can be whipped about by demonic forces into thinking, feeling, and acting in ways that are unholy and unwholesome as well as disappointing to God and themselves. (Read about the perils in trying to separate the power of *Christ* from the identity of *Jesus* in *An Introduction to Volume Two* of *Intelligent Evolution* — in the subsection entitled *The Unfortunate Separation of Christ and Jesus.*)

Using one's imagination is dangerous when that imagination entertains thoughts associated with fear, pride, vanity, willfulness, sexual lust, greed, covetousness, jealousy, envy, hatred, revenge, and unforgiveness. When the imagination entertains thoughts associated with those feelings, the imagination opens up cognitive portals to increased demonic attack. To be sure, all souls in corporeality, by virtue of their being in corporeality, are subject to external temptations. However, their own unholy emotions and desires make them even more susceptible to influences from demonic forces. When unholy thinking is entertained, demons are able to hook their parasitic claws — which is to say, patch their illusions — more easily into the brain's cognitive framework in order to create imagined scenarios that further fan the emotional flames of unholy thinking. If such unbridled feelings are permitted to grow in intensity by our continuing to indulge them, this can place us in significant jeopardy. Entertaining unholy thoughts and feelings holds one's will power and self-control in spiritual abeyance[12] and, thereby, makes the human brain more susceptible to receiving external images from unclean spirits. That is why saved fallen souls in corporeality must be on guard continually and maintain self-discipline at all times.

Heartfelt personal declarations and affirmations should daily include:

1. "I actively, earnestly, and sincerely desire and seek through Christ Jesus to think only the pure and holy thoughts of the Lord God Almighty."

2. "I actively, earnestly, and sincerely desire and seek through Christ Jesus to feel only the pure and holy emotions of the Lord God Almighty."

---

[12] *Abeyance* here means: (1) a state without rightful control or without a rightful owner; (2) a state of being temporarily unoccupied; (3) a state waiting for a claimant; and, *by extension*, (4) a metaphysical state of increased susceptibility to external demonic attack. An example of the word's use in a sentence follows: *When entertaining unholy thoughts and feelings, full property rights concerning the human brain are held in spiritual abeyance until the rightful owner rebukes the unholy thoughts and feelings and repents of all sins committed in connection with those thoughts and feelings — even if the sins are committed only in one's own imagination.*

3. "I actively, earnestly, and sincerely desire and seek through Christ Jesus to express only the pure and holy thoughts and emotions of the Lord God Almighty."

4. "I actively, earnestly, and sincerely desire and seek through Christ Jesus to commit only the pure and holy actions of the Lord God Almighty."

5. "I actively, earnestly, and sincerely desire and seek to be a pure channel of the Lord God Almighty through Christ Jesus."

To summarize at this juncture, when human beings entertain unholy thoughts and feelings, they extend an open invitation to unclean spirits to participate in their mental activities. (Unclean spirits are the disembodied souls of dead people who have consciously rejected Christ while in human form.) The present author knows much about this topic because he was born with a *susceptibility* to receiving external words, ideas, and images (some people might call it a *sensitivity* to receiving impressions). In other words, the present author has the capacity to receive words, ideas, and images from incorporeal sources. (For the sake of clarity, words convey some ideas and images but not all ideas and images.) This susceptibility has worked, at times, to his advantage and, at other times, to his disadvantage. To be sure, like the Apostle Paul, the present author had a thorn in his flesh, or "messenger of Satan" *(2 Corinthians 12:7 KJV),* assigned to him for decades. This angel of Satan was only recently removed (in January 2017) by the Lord Jesus Christ in his grace and mercy. (Although I had known for decades that I was being attacked demonically, I did not know that it specifically was by an angel of Satan. I thought it was by an unclean spirit. I did not learn of the exact nature of the attacks until I was informed by a heavenly source that the thorn was going to be removed from me. As strange as it may sound, about the time of the evil messenger's extraction, we even said good-bye to one another with the understanding that we would never again be in each other's presence.)

Note: Just because a thorn in the flesh has been removed does not mean that one cannot conjure up demonic beings by entertaining sinful desires, thoughts, and scenarios.

In keeping with Aristotle's views on the usefulness of time, time may or may not be an agent of change: Just think of the many people who have lived long lives and who have not matured cognitively, emotionally, spiritually, or socially. Not in keeping with Aristotle's views on time that "time is everywhere the same" *(Physics, IV.14, p. 116)* are Einstein's theories of relativity. Einstein's theory of special relativity states that time slows down or speeds up depending on how fast one is moving in relation to something else, and Einstein's theory of general relativity states that time is bent by gravitational fields. Consequently, in contrast to the thinking of Aristotle, time in the physically-observable universe is everywhere *not* the same. This is an important correction to Aristotle's thinking that opens a window to understand the relativity of time concerning certain aspects of cosmic evolution, biological evolution, and consciousness evolution.

As previously illustrated, students of metaphysics may borrow language from Aristotle and apply it to contexts not intended by Aristotle. *For example,* Aristotle's comment that "a movement is not made up of movements but of discrete changes of place" *(Physics, VI, 1, p. 140)* can be applied to cosmic evolution, biological evolution, and consciousness evolution, especially if *intelligent evolution* is viewed as one sweep of the Creator-God's hand rather than a sequence of discrete actions. In order to understand this view, one needs to understand *the Final Cause* (that is, the traditional teleological cause) of salvation's availability for fallen souls in corporeality: This was always the main reason for any and all ordering and non-randomization in the physically-knowable universe in the first place (as well as in the last place). *As an additional example,* Aristotle's comment that "it takes infinitely many rather than finitely many *nows* to make contact with infinitely many things" *(Physics, VI.2, page 143, italics mine)* can be applied to the eternal separation of the spiritually-observable universe from the physically-observable universe — which is to say, the eternal separation of the *here* and *now* in the spiritually-observable universe from the relative *time* and relative *space* in the physically-observable universe.

The present author has stated that the spiritually-observable universe is dimensionless and that, in it, *now* replaces relative time and *here* replaces relative space. Consequently, things do not move in the spiritually-observable universe in the same way that things move in the physically-observable universe. Dimensional motion requires relative space-time. Aristotle's comment that "nothing moves in the now" *(Physics, VI.3, p. 143)* is in agreement with the present author's assessment, especially if "the now" is the eternal *now.* Further, Aristotle stated: "If it were possible for something to move in the now, there could be both faster and slower motions in it" *(Ibid.)*. Although there are sevenfold divisions in the spiritually-observable universe *(see Revelation 1:4, 3:1, 4:5, and 5:6),* those sevenfold divisions are in the substance, or essence, of Spirit and not measurable in temporal terms. Thus, Aristotle's two statements that: (1) "since there is nothing whose nature is to move in the now, obviously there is nothing whose nature is to rest in the now either," and (2) "the upshot of all this is that the same thing will simultaneously be at rest and in motion," *(Physics, VI.4, 146)* demonstrate compatibility with Eddy's idea that "God rests in action" *(Science and Health 519:25)* as well as to the broader Christian metaphysical concept that there is no such thing as inaction for the Creator-God.

Yes, as recorded in Chapter One of Genesis, the Creator-God did rest on the seventh day from His creative works, but He did not suspend His actions in eternity; He only suspended His actions for a period of time in temporality. Although it may appear in the physically-observable universe that the Creator-God waits, He does not wait in the spiritually-observable universe. This understanding is in agreement with the seemingly contradictory prophetic Biblical teaching that "though the vision of the LORD tarries, wait for it; because it will surely come, it will not tarry" *(Habakkuk 2:3b KJV Paraphrase).*

Divisibility exists in the physically-observable universe but does not exist in the spiritually-observable universe. In contrast, indivisibility exists in the spiritually-observable universe but does not exist in the physically-observable universe. However, indivisibility also exists in the vacuum of infinite space beyond the fringes of the physically-observable universe because, as Aristotle taught, "it is impossible for infinity to consist of finite components" *(Physics, VI.7, page 157)*. Teaching about indivisibility, Aristotle stated: "something that has come into existence has done so at an indivisible moment" *(Physics, VI.5, p. 151)* — in other words, one might conclude that any and all creation through ordering and non-randomization must occur at individual points where the eternal *here* and *now* of the spiritually-observable universe intersect the relative space-time of the physically-observable universe. Thus, in the paradigm of *intelligent evolution,* the Creator-God extended His *hand* — that is, His *action* — from where He resides to where He does not reside (the so-called *elsewhere* of eternity).

Finities can be traversed in finite times. Infinity can never be traversed *(*see *Physics, VI.7, p. 157)*. And traversability does not apply to eternity because all of eternity is *here* and *now* in the spiritually-observable universe. There is neither relative time nor relative space in the spiritually-observable universe. And, as stated earlier by the present author, no one moves in the spiritually-observable universe as we think of motion because there are no dimensions there. Thus, the Creator-God can be referred to as the Prime Mover but only in the physically-knowable universe because nothing moves (as we know motion) in the spiritually-observable universe, the state where He resides. Because "anything moving is moving in time" *(*see *Physics, VI.8, p. 159),* there is no moving where there is no relative space-time. And, although God the Father has sequestered the Glory of His Being in the spiritually-observable universe, He will no longer sequester it when He infuses all that has been redeemed, reclaimed, and restored through Christ Jesus with "the All," or Totality of His Being, at the end of *the Millennium*. To be sure, through the Creator-God, redeemed created beings will then be anywhere in the *here* and *now* that they desire to be.

"Since everything that changes changes in [relative] time and nothing changes in the now" *(Physics, VI.10, p. 165, brackets mine)*, salvation or redemption of the eternal soul is a change that occurs in relative time with ramifications for eternity. Aristotle stated that "no process of change is infinite because (as we have seen) every change… has a starting-point and an end-point" *(Ibid.)*. Thus, one can assume not only that there is both a starting-point and an end-point for the salvation experience while the eternal soul is in corporeality but also that there is neither a beginning nor an ending for salvation in eternity; in eternity, one's salvation simply always *is* as soon as it has been individually received.

Aristotle stated: "Everything that changes must be changed by something" *(Physics, VII.1, p. 167)*. Aristotle's understanding here fits nicely with the thesis of this book that the teleological cause in *intelligent evolution* presupposes the end-result of *Homo sapiens* as the intended corporeal encasement for the fallen eternal soul, specifically providing it with opportunities (in a continuing sense) from the Holy Spirit to be granted repentance and receive salvation.

All evolutionary changes have been made by the Creator-God as the Prime Mover for the intended end-result of *Homo sapiens*. Aristotle stated: "Any immediate agent of change — not in the sense that it is the purpose of the change, but in the sense that it is the original source of the change — is contiguous with what is changed (by 'contiguous' I mean that there is nothing between them)" *(Physics, VII.2, p. 170)*. The Creator-God is not the purpose of the change but, rather, the original source of the change as well as all associated changes. The purpose of all evolutionary change is to provide a suitable habitation for the fallen eternal soul. "That the final agent of alteration and the first object altered are contiguous" *(Physics, VII.2, p. 172)* is as metaphysically true for the teleological cause of *Homo sapiens* as it is for the first protists (protoctists), first prokaryotic cells, and first bacteria. The habitation of the fallen soul in a physical body belonging to *Homo sapiens* is intended primarily for the opportunity of that soul to receive salvation.

Mortality is an altered state of immortality as a result of iniquity and sin. In reverse, immortality is an altered state of mortality for saved fallen souls as a result of their repentance and conversion in conjunction with their acceptance of God's *only-begotten* Son, Christ Jesus. This metaphysical truth provides a lens to understand the most significant change possible for fallen eternal souls. The present author knows this truth because "when a particular appears, the knower somehow knows the universal by means of the particular" *(Physics, VII.4, p. 177)*. Knowing the larger truth from a particular (i.e., induction) is apprehending it *a priori*. This principle is supported by pure reason as well as empirical evidence. (There is more about this in Section 2.5.2 — entitled *Insights, Implications, and Applications from Kant*.)

Aristotle drew from Empedocles when he wrote that things "are changing whenever love is creating a unity out of a plurality or hatred is creating a plurality out of a unity" *(Physics, VIII.1, pp. 185-186)*. Although Empedocles and Aristotle viewed these two changes as cyclic and repeating, the present author proposes what is in keeping with his view on *intelligent evolution* that there can only be one *Big Bang* (that is, only one outward manifestation of iniquity) and only one Infusion of the Totality of the Creator-God's *Supreme Being* in the formation of the All-in-all at the end (i.e., expunction) of all relative space-time. However, the present author agrees with the Aristotelian view that "for each kind of change, there must be things with the capacity for that change" *(Physics,*

VIII.1, p. 186). For example, unfallen immortal beings were created with a capacity to fall, and fallen immortal beings who have not continued to fall toward a *second death* by irrevocably rejecting Christ Jesus still retain a capacity for repentance, conversion, and salvation.

In his assessment of what Plato believed, Aristotle described what many Platonists (as well as Neoplatonists) still believe: "that time has an origin" *(Physics, VIII.1, p. 188).* Not only does the present author also believe that relative time has an origin, the present author believes — based on his understanding of the Holy Bible — that relative time has an end. Whereas time in the physically-observable universe could be viewed as a succession of *nows* (or sequence of related events), time in the spiritually-observable universe can only be viewed as one eternal *now.* In other words, relative time in the physically-observable universe is divisible, and absolute time in the spiritually-observable universe is indivisible. And relative time in the physically-observable universe has a beginning and an end, but absolute time in the spiritually-observable universe has no beginning and no end. The counterpart to *relative time* in the spiritually-observable universe is *eternity* (or absolute time), which is the *eternal now* (that is, *the eternal moment)* just as the counterpart to *relative space* in the spiritually-observable universe is the *eternal here,* which is the metaphysical center and circumference of Heaven.

Aristotle ventured into Plato's, Kant's, and Eddy's intellectual territories when he wrote: "For if it is really true, as some people claim, that being is infinite and unchanging, it remains the case that this is not what our senses tell us and that many things do seem to change" *(Physics, VIII.4, p. 194).* Further, Aristotle speculated that perhaps change is related to a "false belief" *(Ibid.)* — something with which Plato and Eddy would readily agree.

As a side note here, the present author believes that Aristotle's pantheism prevented him from seeing two realities at the same time, one physical and the other spiritual. If Aristotle had not adopted such a narrow view, he might have been able to see that *Being* exists in immortality at the same time that *being* exists in mortality. Similarly, the present author thinks that Eddy's own peculiar Neoplatonism and immaterialism prevented her from seeing two realities at the same time. If she had not adopted her unwavering view, Eddy, too, might have been able to concede that *Being* exists in immortality at the same time that *being* exists in mortality. Requisite to such a view is an unwavering commitment to the sacrificial atonement of Christ Jesus at the same time that one is open to metaphysical theory. Without such commitment and openness, one cannot see two things at once in a stereoscopic union.

Aristotle understood that any change requires: (1) an object that has the capacity to change, (2) a change agent, and (3) an instrument by which means the change agent causes change *(Physics, VIII.5, p. 202)*. Of course, Aristotle did not understand: (1) that Lucifer was the change agent and temptation was Lucifer's instrument for the fall of eternal souls from immortality to mortality; or (2) that Christ Jesus is the change agent and the shed blood of Christ Jesus is the instrument for the return of eternal souls from mortality to immorality.

Aristotle postulated that a first agent of change must itself be unchanging. Although Aristotle was a pantheist, he acknowledged that this first agent of change must be *Deity* (also referred to by Aristotle as *the Prime Mover, the Primary Mover, the First Mover,* and *the First Principle).* Perhaps the so-called *immortal* Greek and Roman gods influenced Aristotle's pantheistic and narrow views on *Deity.* Perhaps Aristotle's subconscious belief in the supremacy of matter influenced such views as well.

Aristotle did not know that the eternal first agent of change is the eternal last agent of change, and that these are the same agent — which is to say, the Creator-God: the divine Intelligence and Supraconsciousness of *the Whole Universe* as well as the God of the Holy Bible. Aristotle's conclusion that "the primary kind of change is movement, that is, change of place" *(Physics, VIII.7, p. 212)* unwittingly provides Christian metaphysicians with insight that the first change in immortals who fell because of their newly-developed iniquity was, in fact, a type of *movement,* or *change of place,* from the state of immortality to the state of mortality.

Aristotle also unwittingly provided language for the Christian metaphysician to describe the ascendancy of the saved eternal soul as it figuratively wafts in return to its Creator-God: "Anything which is coming into being is incomplete, and is in progress towards its cause" *(Physics, VIII.7, p. 212)*. And to those who might erroneously conclude that Good and Evil are in unity in the scheme of things, Aristotle's language has great bearing on the separation of the two when he wrote: "Opposites, however, are different in species and do not constitute a unity; and the distinctions mentioned are differentiae of place" *(Physics, VIII.7, p. 217)*. In other words, Good and Evil are separate and in different places. This provides a solid argument against the erroneous conclusion in Daoist metaphysics that a yin and a yang co-exist harmoniously in the same oneness reality.

The last summary statement that the present author shall use from Aristotle's *Physics* before he turns to Aristotle's *Metaphysics* is that "the eternal first agent of change has no magnitude, and is located at the outer edge of the universe" *(Physics, VIII.10, p. 227)*. This language is in total agreement with what the present author has proposed concerning the dimensionless nature of eternity and *Deity* and the representation of the physically-observable universe's relationship to the spiritually-observable universe, which relationship is depicted in Figure One of this book (on page I-23).

2.5.1.2 Aristotle's *The Metaphysics*

Aristotle's *Metaphysics* is definitely a book for thinkers. As with his *Physics,* it is important to inform readers and listeners that the present author has applied concepts and terminology from

Metaphysics in somewhat different contexts than intended by Aristotle. In other words, the present author has used Aristotelian concepts and language labels that have utility for a discussion on *intelligent evolution.*

Cause, or what might be thought of as *reason for existence,* is very important to Aristotle's *Metaphysics.* Aristotle distinguished between the skilled person and the "merely experienced" person based on one's knowledge of *cause.* He stated that "the skilled know the cause, whereas the experienced do not" *(Metaphysics, Alpha 1, p. 5).* From this, Aristotle deduced that "the skilled can, whereas the merely experienced cannot, teach" *(Ibid.).* Applying this to the paradigm of *intelligent evolution,* it is the skilled person, not merely the fact-based person, who can put his or her understanding of *cause* to work for elucidating the Creator-God's teleological cause of *Homo sapiens* as the reason for all evolutionary change culminating in the emergence, appearance, and physical form of that species. Aristotle concluded that "wisdom is knowledge having to do with certain principles and causes" *(Metaphysics, Alpha 1, p. 6).* According to the present author, wisdom for a human being is the ability to *see through* to the spiritually- or metaphysically-observable universe while one is still in the physically-observable universe. And without the Creator-God's Holy Spirit residing within one's soul, it is impossible to *see through* to obtain spiritual wisdom. *Seeing through* to eternity also requires doing the Will of the Creator-God. Such *seeing through* is spiritual sight unimpeded by carnal consciousness.

Aristotle posited, and the present author concurs, that "theoretical knowledge is more capable of teaching [about the science of] causes" *(Metaphysics, Alpha 2, p. 8, brackets mine).* Aristotle then wrote what Eddy could have penned: (1) "For this science must be theoretical of the primary principles and causes" *(Metaphysics, Alpha 2, p. 8);* and (2) "such a science… would be that which a god would most choose [because] that is the one of the sciences that is divine" *(Metaphysics, Alpha 2, p. 9, brackets mine).* Concerning this topic, Aristotle concluded: "We have, then, said what the nature is of the science that we are seeking and what the end is at which the search should aim and the whole method" *(Metaphysics, Alpha 2, p. 10)*. In other words, according to Aristotle, the nature of the science, or philosophy, sought is *divine* (i.e., *divine science*) — of which *First Philosophy*, or *metaphysics*, is the most important. According to Aristotle, *First Philosophy,* or metaphysics, is the study of *First Principle,* or God. (Thus, *First Philosophy* is also theology.)

Concerning the notion of *intelligent evolution,* although the capacity for change is found in living substance (in this case, protoplasm), "it cannot be that the substrate forces itself to change" *(Metaphysics, Alpha 3, p. 14)*. Aristotle acknowledged the belief that "mind was present in the universe, as in the animals, and that this was the cause of order in nature" *(Ibid., p. 15)*. From these two preceding statements, we gain greater insight into the nature of the teleological cause for *Homo sapiens.*

In the fourth section of his first discourse in *Metaphysics,* Aristotle acknowledged Empedocles' contribution to philosophy that "love is the cause of all good things and strife [is the cause] of bad things" — which therefore implies that "good itself is the cause of all good things" *(Metaphysics, Alpha 4, p. 17, brackets mine)*. Aristotle also acknowledged Anaxagoras' contribution of "mind as a device for the making of the cosmos" *(Ibid.)* — hence the conclusion that both the presiding principle and teleological cause of the entire cosmos are in *a universal Mind.*

Over the past two millennia, philosophy has been especially undervalued by many Christian people. Aristotle stated that "the study of truth is called philosophy" and "truth is the aim of theoretical thought as action is of practical thought" *(Metaphysics, Alpha the Lesser.1, p. 44)*. His comment that "we do not know the truth without [knowing] the cause" *(Ibid., brackets mine)* applies directly to religious and spiritual living because human beings cannot know the truth without first knowing the *Primary Cause.* Human beings must come to recognize the one true and only real

Creator-God as the *Primary Cause* as well as the *First and Final Cause*. Christian people must never be afraid to search for depth in meaning behind the truths of the Holy Bible.

Philosophy should be important to all Christian people in order to have discussions about moral and ethical issues. It is too simplistic to castigate certain acts as representative of moral relativism and, therefore, as "non-Christian" when one has not thought thoroughly through the reasoning behind the acts. *For example,* to believe that all murder is wrong except for assassinating someone like Adolf Hitler ideally requires dialectical contributions from both sides of such an equation *before* a decision is made to endorse such an act, refrain from endorsing it, condemn it, or refrain from condemning it.

When Pontius Pilate asked Christ Jesus if he were a king, Christ Jesus responded: "To this end was I born, and for this cause came I into the world, that I should bear witness to the truth," and "everyone that is of the truth hears my voice" *(John 18:37 KJV Paraphrase)*. Because he was not grounded in Messianic expectancy, Pontius Pilate could only respond with the rhetorical question "What is truth?" *(John 18:38 KJV)* Like all other people who have not hoped for Christ, waited for Christ, recognized Christ, and accepted Christ, Pilate could not recognize truth enough to trust in it. For Pilate and so many others like him, there can only be philosophical conundrums when Christ Jesus is not known as Savior to them. Without Christ Jesus, people are "ever learning but never able to be brought to the knowledge of the truth" *(2 Timothy 3:7 KJV Paraphrase)*. Theoretical expeditions like *Intelligent Evolution* can only be fruitful if one is grounded in the knowledge of Christ Jesus as Savior, *only-begotten* Son of God, and God Incarnate.

If one is grounded in the knowledge of Christ Jesus, truth new to the learner should be sought — not truth that is contradictory to the truth in the Holy Bible, but truth that is complementary to it. In order to learn new truth, one must regress[13] to the primary cause, the first cause, the eternal cause, the teleological cause, and the final cause (all one and the same), knowing the Creator-God Himself by understanding who He is through what He is doing.

That Christ Jesus is referred to in the Holy Bible as the *Logos* has significantly more meanings in addition to *Word, Principle, Thought,* and *Speech* because the Greek verb from which *Logos* is derived not only means "to speak or say" but also "to collect or gather." Christ Jesus is the spoken Word of the Creator-God that not only creates but also *gathers together* that which was fallen and lost. Everyone who accepts Christ Jesus as Redeemer is re-created, born anew, and gathered into the Body of Christ. In other words, as Christ Jesus re-creates in salvation, he *gathers* in, or harvests, fallen souls who are no longer lost because they are now saved.

Regressing to the primary cause, the first cause, the eternal cause, the teleological cause, and the final cause (all one and the same), we bend back in the continuum of relative space-time to understand the Creator-God as Prime Mover, Primary Existent One, and Divine Cause. Aristotle pronounced truth when he stated that "it is impossible that the primary existent, being eternal, [can] be destroyed" *(Metaphysics, Alpha the Lesser.2, p. 46, brackets mine),* and "we are thought to know when we have cognition of the causes" *(Ibid., p. 47)*. Relating this to the paradigm of *intelligent evolution,* we can only really know something when we learn to know Christ Jesus, who is the *Cause* of all-that-is. And we are led to trust the conclusion that, because the Creator-God cannot be destroyed, then the Destroyer (Satan or the fallen Lucifer) can only destroy himself and those who belong to his destruction. Conversely, if "the created arise from the creator" *(Ibid., p. 46),*

---

[13] **Throughout this book, *regress* means "to reason backward," *regressing* means "reasoning backward," and *regression* means "the act of reasoning backward." *Regress, regressing,* and *regression* are related to *teleology* by working backward from all that currently exists to the *First and Final Cause* of everything that is Good: the creative *Logos,* divine *Principle,* and spoken *Word.***

then all who have returned to the Creator-God through Christ Jesus cannot be destroyed. If the reader of *Intelligent Evolution* understands what is posited in this book, then he or she is becoming a *knower* of truth because, as stated by Aristotle, "the learner is the *becoming* knower" *(Ibid., italics mine)*. "For now we [who are in corporeality] see through a glass darkly; but then [when we are fully restored to immortality, we shall see the truth of Christ Jesus] face to face: now I know in part; but then shall I know even as also I am known" *(1 Corinthians 13:12 KJV Paraphrase, brackets mine)*.

Aristotle stated that *the science of substance,* or *the science of essence,* is the science of the primary causes, against which all other sciences are to be measured. In other words, all other sciences are of lesser value than *metaphysics,* or divine science. And, in that "all actions involve movement" *(Metaphysics, Beta.2, p. 60)*, we may deduce that all cosmic, biological, and consciousness evolutionary changes, regardless of other factors, constitute metaphysical movement from the actions of the Creator-God, the Primary Mover, across the skeins of relative time.

Aristotle acknowledged that the absurdities in anthropomorphism are illustrated when people believe that, of sensible objects (i.e., physical objects), "some are eternal and some perishable." He stated that, in doing so, people are "making the same sort of mistake as those who say that there are gods but that they are in the form of men: For they are doing nothing else than positing eternal men [i.e., eternal human beings], and these thinkers are not positing forms but eternal sensibles [i.e., physical beings]" *(Metaphysics, Beta.2, pp. 62-63, brackets mine)*. In other words, the concept of *immortal mortals* is as absurd as the concept that the essence of ideas is perceived by the physical senses.

As you, the reader or listener, will come to understand ever more fully, all five thinkers covered in *Intelligent Evolution* exercised rational argumentation. Therefore, in that all five of them gave evidence of superior thinking by using it, rational argumentation is one of the techniques that you should be looking to develop from studying this book — which techniques include induction, deduction, regression, logic, syllogism, affirmation, and refutation to lay the groundwork for conceptual understanding.

Using the language of Aristotle, one should look to lay the groundwork for the conceptual understanding of *insensible objects* through *extra-sensory, supra-sensible,* or *hyper-sensible* perception. To be sure, such searching is more Platonic than Aristotelian. And, although rational argumentation can help you to gain insights, implications, and applications, rational argumentation cannot convince other people of anything unless the Creator-God's Holy Spirit convicts them of its validity (when they are ready and willing to be convicted).

Without having studied the Tanakh (the Hebrew Bible), Aristotle was able to say that "those who do not eat the nectar and ambrosia [of the gods] are born mortal" *(Metaphysics, Beta.4, p. 68, brackets mine)*. To be sure, Aristotle's understanding is consistent with the Genesis account of the fall of Adam and Eve from immortality to mortality and the Creator-God's prohibition of their eating fruit from *the tree of Life* in Eden. Aristotle's recognition of this truth is more than just coincidence. Aristotle's metaphysical insight illustrates that truth is seen as truth by those who are open to seeing it. For the sake of clarification here, the *food* of immortals is never the food of mortals and, conversely, the food of mortals is never the *food* of immortals. (The primary reason that Christ Jesus ate after his resurrection was to demonstrate to his followers that he was not a ghost, spirit, or illusion but, instead, someone whose corporeal body had been raised from the dead fully functioning.)

Aristotle asked the question "How is it that from imperishable things perishable things should come?" *(Metaphysics, Beta.4, p. 70)* He added: "But indeed if unity itself and being itself exist, then there is a great puzzle how there will be anything apart from them" *(Ibid., p. 71)*. Like Eddy, Aristotle ignored the effects of an Adamic Fall. In Aristotle's case, he was certainly aware of the mythological effects from opening Pandora's box, but he simply had not been properly educated about the Adamic Fall from a Biblical perspective. In contrast, Eddy chose to ignore the Adamic Fall because, in the present author's estimation, that was her greatest insult to its cause as well as to its effects, and because she believed that evidence of the Fall could not exist in the spiritually- or metaphysically-observable universe (using the present author's terminology), especially if one viewed it as unaltered from the effects of Satan, or, in Eddy's words, *mortal mind.* To be sure, Eddy would have agreed with Aristotle's quote of Parmenides that "it is necessary that all things are one and that this is *being" (Ibid., italics mine).* From the present author's perspective, all questions are eventually answered, all puzzles are eventually solved, and all discrepancies are eventually reconciled when we personally know the Creator-God through Christ Jesus, the *Cause* of all-that-is. That Aristotle described the role of potentiality in relationship to cause allows the inquiring Christian to extrapolate that the potentiality of souls to receive salvation is the sole teleological cause, reason, and purpose for the biological evolution of *Homo sapiens:* The species *Homo sapiens* exists to provide opportunities for salvation and eternal redemption to souls fallen from immortality to mortality.

According to Aristotle, the science of being is *First Philosophy,* the study of the causes and principles of *being as being* ("being *qua* being") — which expression is just another way of referring to "*being* by virtue of itself" and "*being* of, in, and through itself." Aristotle stated that "the science [i.e., First Philosophy] we have specified must also cognize the opposites of the things that we have mentioned, the *other,* the *dissimilar* and the *unequal* and such other things as are spoken of either in relation to one of these or in relation to plurality and the *one" (Metaphysics, Gamma.2, p. 83, italics and brackets mine).* We can infer from Aristotle's position, then, that Eddy actually substantiated the existence of *mortal mind, mortal man,* and *error* because she argued against their existence and because she invented her own language labels for them: One does not name and give methods for combating what does not exist. *For example,* if the present author argued against the existence of Evil, the present author would actually give assent to its existence even though he wished to demonstrate its non-existence. In truth, in doing so, the present author really would want to argue that there is a power *greater* than the power of Evil — even though he was seeking to negate its potency. In other words, when we say that Evil does not exist, we are actually cognizing the opposite of Good (that is, Evil) as we try to treat its effects. Thus, when arguing that mortal mind, mortal man, and error do not exist, we substantiate that they do exist even if we do not wish to grant them the status of *being*. Aristotle was correct in claiming that "*it is not possible to say truly at the same time that the same thing both is and is not a man" (Metaphysics, Gamma.4, p. 92)*. In application, saved human beings cannot be immortal beings and mortal beings at the same time; they must be one or the other. (It is the position of the present author that all saved human beings are restored immortal beings when they are saved even though they are still in corporeality.) Likewise, unsaved human beings cannot be immortal beings and mortal beings at the same time; they must be one or the other. (It is the position of the present author that all unsaved human beings remain mortal beings.) To be sure, all souls are eternal regardless if they are immortal or mortal.

It is somewhat ironic that arguing against, refuting, or denying the existence of something actually affirms its existence. Casting out demons, healing people of illnesses, and quelling stormy seas, Christ Jesus offered no argument against their existence. Christ Jesus did not refute or deny that negative conditions existed. Rather, he affirmed the existence of Evil, sickness, and inclement weather as he rebuked them. Christ Jesus did not heal people of illnesses that did not exist; there would have been nothing miraculous in doing that. In truth, rather than denying the existence of Evil and sickness, Christ Jesus affirmed not only that they existed but that there also existed a power that had absolute sway over Evil and sickness. Because the Creator-God was, and is, the source of his power, and because he was "God in the flesh" *(1 Timothy 3:16 KJV Paraphrase),* Christ Jesus used his power to command wholeness, health, and peace into existence. That Christ Jesus said that Jairus' daughter was only sleeping when everyone knew that she was already dead was done for the purpose of teaching human beings that they should "have no confidence in the flesh" *(Philemon 3:3 KJV)* and, therefore, not trust appearances. In other words, there is a reality greater than corporeality and the maladies to which it subscribes.

Concerning natural science (more properly called *natural philosophy* for Aristotle's time), Aristotle stated that "there is a science higher than natural science" and "natural science is a kind of philosophy, but it is not *First Philosophy*" *(Metaphysics, Gamma.3, p. 87, italics mine)*. In other words, Aristotle taught that metaphysics is of primary importance and that natural science, or natural philosophy, is less important than metaphysics. Although the reader of, or listener to, *Intelligent Evolution* needs to understand natural science (just as the geologist Charles Lyell and the naturalist Charles Darwin gathered facts and made observations on nature and natural history), the reader or listener primarily needs to understand metaphysics in order to comprehend the spiritual First Cause, or First Principle, of *intelligent evolution* in order to make complete sense of their dependent physical causes and principles that gave pulse to the various stages of cosmic evolution, biological evolution, and consciousness evolution.

In most instances, Aristotle wrote about the perception of *sensible things* (that is, physical objects) and not about the discernment of spiritual things (that is, intangible objects). The present author believes that there really was no dichotomy — and, therefore, no real dilemma — for Aristotle between the visible and invisible realms (at least none that Aristotle would directly acknowledge). Modern students of Aristotle's writings must be careful not to attribute to Aristotle, or read into his writings, contemporary understanding from the thinking of others that occurred subsequent to Aristotle's death. Although the present author has read *beyond* what Aristotle wrote in order to apply Aristotle's concepts and language labels to contexts other than those intended by Aristotle (in particular, to the paradigm of *intelligent evolution*), the present author has not attributed to Aristotle what Aristotle himself did not understand, have knowledge of, or wrote about.

The following two statements are not contradictory to one another: (1) The Creator-God created everything that exists (excluding Evil, demons, iniquity, and sin). And (2) Logos-driven evolution is responsible for cosmic, biological, and consciousness developments across all relative time and relative space. In addition to *Supreme Being,* the hyphenated word *Creator-God* describes a self-existent force and cause. But the word *evolution* does not describe a force or cause. *Evolution* is a noun that describes a process impelled by a force or cause, but evolution, in itself, is not a force or cause. In fact, evolution is a process that has been shepherded, one step at a time, by the Great Shepherd. Evolution did not guide itself. Evolution has no consciousness of its own although it was designed by *consciousness-in-itself,* the Supraconsciousness of the Creator-God.

Metaphysical language is the language of ideas. Regardless if the Creator-God's Holy Spirit uses words, images, or actions to speak to us, the Creator-God communicates to all of His created, both in Heaven and on Earth, using ideas. When the Creator-God shares His ideas with people on Earth, hopefully they listen and ponder. And when we share our ideas with the Creator-God, His amusement is kindled because our ideas are so deficient in comparison to what is seen and understood by His people in Heaven. The Creator-God has spoken His ideas into existence in the physically-observable universe through the process of *intelligent evolution*. That is why His people on Earth need Christian metaphysics to understand the cause, reason, purpose, and principle behind the sweep of His hand and the utterance of His mouth. Without thinking metaphysically, it would be impossible for us to understand the intricacies of what the Creator-God has created and made. To be sure, when we are genuinely thinking intelligently in accord with His Will, we are reflecting the Creator-God's complete image and perfect likeness.

Aristotle acknowledged that both *presence* and *privation* "are causes as sources of process" *(Metaphysics, Delta.2, p. 116)*. This idea certainly has application to the process of *intelligent evolution* in which the presence or absence of multiple factors impacts significantly on natural selection in both microevolution as well as macroevolution (see Footnotes 2 and 3 in Volume One). Aristotle added that "a principle of change or process in this way is said to be a *potentiality*" *(Metaphysics, Delta.12, p. 131)*. Applying this to the paradigm of *intelligent evolution,* we can unequivocally state that every cell has the potential to become any other cell over relative time and relative space. Thus, from a metaphysical standpoint, not only is a stem cell totipotent in an individual (depending on conditions present and/or absent, of course) but also each cell is totipotent on an evolutionary scale of change (again, depending on the conditions present and/or absent).

Biologists understand that viruses cannot be changed into cells because viruses are not now, nor have they ever been, cells (perhaps they were once parts of cells, but they were never entire cells). Aristotle would explain that this "non-potentiality [of viruses] is a privation of potentiality" *(Ibid., p. 133, brackets mine)*. (The present author again reminds the reader or listener that he is simply borrowing concepts and language labels to help further explain metaphysically what is understood empirically today — in this far future after Aristotle's death.) Although some viruses and bacteria have emerged after the origin of *Homo sapiens,* no virus or bacterium has replaced *Homo sapiens* as the last teleological rung on the evolutionary ladder.

The present author concurs with Aristotle that *First Philosophy (First Science),* or *metaphysics*, is really *theology* in addition to *wisdom*. If only Christians could look at their own personal theologies as philosophies based on their understanding of the Holy Bible, they would have a much easier time discussing with others the significance of what they hold to be true and what they hold not to be true. They would feel less threatened and be better able to dialogue with others about their personal belief systems:

> For who can doubt that if there is Divinity anywhere in the universe, then it is in the nature studied by First Science that It [sic] is to be found. And it is also for [this] Supreme Science to study the Supreme Genus [Deity]. And contemplative study is to be chosen above all other sciences, but it is this First Science of Theology that we must prefer to all other kinds [of science, including mathematics and natural science] even [above] contemplation. [brackets mine]
>
> *Metaphysics, Epsilon.1, p. 155*

The crux of the dilemma between natural science and metaphysics is articulated in Aristotle's *Metaphysics* when he posited the following for himself as well as for his readers:

> Either (a) there is no other substance beyond those furnished by nature, in which case the science of nature [i.e., natural science] is the First Science, or (b) there is some Substance that is without change, and, if (b) is true, then that Substance is prior to all others and the science of it is First Philosophy [i.e., metaphysics] — and such a science is universal just because it is first. [brackets mine]
>
> *Metaphysics, Epsilon.1, p. 156*

Do accidents happen in *intelligent evolution?* Yes, but they are not determinants of the emergence, survival, sustainability, or thrivability of the teleological cause or end. To be sure, one should ask a different question: "If the outcomes of chance are foreknown by the Creator-God, do accidents really happen?" The present author believes that Aristotle would weigh in on that question as follows: because "there is no science of [the accidental]" *(Metaphysics, Epsilon.2, p. 160),* accidents really cannot be studied as a whole but, rather, one at a time and, as a result, their impacts can only be studied one at a time. Thus, there is no science to accidents in *intelligent evolution.*

Only the mind that is in Christ Jesus is able to distinguish between and among: (1) what is really true, (2) what we would like to be true but is not, and (3) what is unreal but appears to be true. As indicated previously in *Intelligent Evolution,* Aristotle understood that an affection of the mind is something that the mind only desires to be true and, therefore, has a predisposition, or bias, for it to be true. Without Christ Jesus in our lives as Savior and Sovereign King, we are untethered enough to easily convince ourselves that something is true because we would like it to be true or because it is presented to our senses and sensibilities as true. For these reasons, we must be careful in our exploration of the spiritually- or metaphysically-observable universe not to project onto it our own interpretations of it or accept someone else's interpretations of it. We must try to remain objective even though our own language labels inject subjectivity into how we describe what we discern and apprehend.

Aristotle asked many questions that are germane to our discussion of *intelligent evolution.* Two of these questions include: (1) "Are there, or are there not, any substances besides the perceptible ones?" and, if so, (2) "What is [their] mode of being?" *(Metaphysics, Zeta.2, pp. 171-172, brackets mine)* For the physically-observable universe, "it is matter that turns out to be [its] substance" *(Metaphysics, Zeta.3, p. 175, brackets mine).* Although Aristotle's writings are sometimes obtuse, pedantic, rambling, and tedious; and although Aristotle sometimes used trivial examples, the student of metaphysics can still dissect out one very important truth from them — which is that *substance* and *essence* are synonymous. Thus, for corporeality, the essence of the physically-observable universe is its substance, matter; and, thus, for immortality, the essence of the spiritually-observable universe is its substance, Spirit. So, the answer to Aristotle's first question posed at the beginning of this paragraph is "Yes, there are substances besides perceptible, or sensible, ones." And the answer to his second question posed at the beginning of this paragraph is "Spirit is the mode of being, or substance and essence, of the physically imperceptible substances in immortality and eternity."

Because the souls of saved fallen created beings have become immortal again (although souls lost their immortality at the time of the Adamic Fall, they never lost their eternality), such immortal beings are able to come to know what the unsaved cannot know as long as they remain unsaved. (All unsaved have the potential to be saved unless they are already beyond reclamation because they have blasphemed the Creator-God's Holy Spirit by saying that the Holy Spirit is the author of Evil's lies.)

Although a thing is not identical with its cause, "a thing must be identical with its essence" *(Metaphysics, Zeta.6, p. 185)*. Thus, although saved souls are not the Creator-God (even though they have been re-made in His complete image and perfect likeness), they are nevertheless identical with His essence — meaning, they are one with Him and in Him through His Holy Spirit. So, although saved fallen souls will never become the Creator-God, they again possess the same essence as the Creator-God and are one with Him in this way. It is the hope of the present author that the religious/spiritual reader or listener will clearly understand the benefits from using philosophy to answer difficult theological questions such as "Upon believing in Christ Jesus, how is it that we actually become one with the Creator-God?" (The answer is "We become one with the Creator-God through His substance or essence — which is to say, His Holy Spirit.")

Aristotle's doctrine of immanent form in matter is in stark contrast to Plato's theory of transcendent Forms in intelligible substance (or *Spirit)*. For Plato, *Forms* are non-material and abstract — nonetheless substantial — ideas that constitute the essences of physical objects and physical qualities. Plato's *Forms* provide the essential bases of our one true and only real spiritual reality. Physical objects may be grasped by one's hand, but Forms can only be grasped by one's imagination as ideas. However, *Forms* are not products of one's mind. They are not products of one's mind because they have objective reality. For the present author, the spiritually- or metaphysically-observable universe is the world of Plato's *Forms*. For Aristotle, because of his pantheism, "it is patently the case that there are no Forms" *(Metaphysics, Kappa.1, p. 318)*.

Although Aristotle did not believe in the existence of *Forms,* he did believe in the existence of *forms* (lower case *f*). Again, Plato's *Forms* are in contrast to Aristotle's "forms," which are their physical counterparts. Both Plato and Eddy would say — if they used the same language labels — that physical *forms* are not real because they are merely outward appearances that belong to a shadow or crepuscular world rather than an ideal world. (Regardless of these positions, the present author was told by someone in Heaven that *Life is precious in all its forms* — which, I assume, includes *everywhere Life is found.*)

For Aristotle, there is only a sensible, or sensory, world. For Plato, there is only an intelligible extra-sensory, supra-sensible, or hyper-sensible world connected to a corporeally sensible, or sensory, world that is an illusion. In opposition to Plato, Aristotle would claim that because *Forms* do not exist independently, they must be non-existent. Whereas Plato would claim that the evidence for *Forms* is intuitive based on the memory of a soul prior to its birth in human form, Aristotle would claim that the evidence for "forms" [in this case, physical appearances] is rational and "based on principles of demonstrative reason" *(Metaphysics, Kappa.1, p. 317)*. For Plato, the Creator-God is transcendent, which means, in the language of the present author, that He is outside of the physically-observable universe. For Aristotle, the Creator-God is immanent, which means that He permeates the physically-observable universe. Unfortunately, people who create dichotomies like Plato and Aristotle are unable to comprehend that the following two statements do not require mutual exclusivity: (1) *God transcends physical nature.* And (2) *God is immanent in physical nature.* To be sure, it is in reconciling and blending these two together that permit one to understand the conceptual framework not only for *intelligent design* but also for *intelligent evolution.* The present author blends the previous two statements into the following capstone statement: *God transcends physical nature at the same time that He is immanent in physical nature through His intelligent design and evolution of it.*

For Aristotle, "the definition [of something] is an *account" (Metaphysics, Zeta.10, p. 201, brackets mine)*. This provides a useful tool to help us define the Creator-God through the account of Him that we find in the Holy Bible. An account of what the Creator-God has done, is doing, and will do helps us to define Him; and, by defining the Creator-God, we get to know Him. The Creator-God is not indefinable. In fact, all saved people personally know the Creator-God by knowing Christ Jesus, their Savior through the blood he shed at Calvary.

In stating that "some matter is perceptible but some intelligible" *(Metaphysics, Zeta.11, p. 208),* Aristotle laid the conceptual groundwork for understanding an *unseen invisible* of physical nature and matter vis-à-vis the atoms of elements. Although Aristotle did not discover the atom, he speculated as to its existence.

In many instances, human beings need to be able to conceptualize what it is they are looking for in order to find it. This was certainly true for discovering the atoms of elements as well as their subatomic particles. And it is true for discovering initial events in *the Big Bang* that occurred almost fourteen billion years ago during the formation of the physically-knowable universe. Especially applicable here is Aristotle's statement: "This is in any case part of our purpose in trying to frame definitions for perceptible substances. After all, it is really up to physics and Second Philosophy to give us a theory of perceptible substances" *(Metaphysics, Zeta.11, p. 209)*. In the words of the present author, although it is up to Christian metaphysics to give us theories of cosmic evolution, biological evolution, and consciousness evolution, it is not up to Christian metaphysics to delineate the roles that specific physical forces play in elemental, chemical, atomic, and subatomic interactions. Metaphysicians leave that delineation to physicists and, in the words of Aristotle, *Second Philosophy.*

Although Aristotle did not consciously imply the existence of an independent world of Spirit, we can infer from his writings that he believed that the world of Spirit existed, just not independently of matter. Students of Aristotle can conclude that Aristotle believed in the fusion of the *invisibly unseen* to the *visibly seen.* In other words, Aristotle believed in *composites* of form and matter. For *example,* Aristotle believed that the human soul always manifests as flesh and blood in a "conjunction of body and soul" *(Metaphysics, Theta.7, p. 271).* Aristotle did not conceptualize the human body and the soul separately. Because Aristotle did not perceive them to be separate, he was unable to conceptualize that souls also exist independently of corporeality and matter.

Although Aristotle did not believe that matter is the primary substance, he did believe that "there can be no doubt that [matter] is a substance" *(Metaphysics, Eta.1, p. 234, brackets mine)* and "the hallmark of all perceptible substances is the possession of matter" *(Ibid.).* Students of Aristotle can only conclude that he thought of substance, or essence, as a composite of matter and its physical attributes, including "form."

Aristotle's comparison and contrast of potentiality and actuality is germane to the study of *intelligent evolution.* Important related concepts reveal that: (1) physical things change because they have the capacity to change; (2) when physical things change, it is because they have come in contact with an agent of change; and (3) when changed, physical things demonstrate their actuality. The application of these truths to *intelligent evolution* is that potentiality exists in animate beings to provide the platform for something else — specifically, other animate beings from genetic changes. That animate beings change over relative time and across relative space has resulted in biological evolution and consciousness evolution. Aristotle stated: "About such things, we can make a generalization: in all cases such a thing is potentially the next item in the series" *(Metaphysics, Theta.7, p. 270)*. The evolution of species, however, does not mean that one species begins where another species ends. If that were true, there would only be members of final evolutionary events (i.e., the most recently emergent species) and no other species that helped give rise to them would be extant.

That "the end is *the actuality*" is true for *Homo sapiens* because *Homo sapiens* has always been the Creator-God's desired physical end for cosmic, biological, and consciousness evolution. But there is a subsequent *actuality* that is even more desirable and that is the re-immortalization of the eternal soul through the shed blood of Christ Jesus, which is the Creator-God's desired spiritual end for biological evolution. In other words, human beings are not truly actualized (or re-immortalized) until they accept salvation through Christ Jesus. The shed blood of Christ Jesus provides the mechanism and instrument of *entelechy* (i.e., the realization of potential) for the eternal soul's change from mortal to immortal, the *immortal* here both the soul's original, or pre-fallen, form and the soul's final, or post-fallen, form when saved.

In his *Metaphysics,* Aristotle stated: "I think we have made the point: actuality has priority not only over potentiality but over every principle of process" *(Metaphysics, Theta.1, p. 277)*. The conclusion of the present author is that understanding potentiality and actuality expands individual human consciousness.
For Aristotle, *a being* is either inanimate or animate. For the present author, (1) *a being* is a living thing that has intelligence, consciousness, self-awareness, and free will; (2) *a human being* is a living thing with the characteristics just given that belongs to the genus and species of *Homo sapiens;* (3) *an immortal being* is a spiritual being whose eternal soul is in Heaven with the Creator-God; (4) *an immortal soul* is one that has been saved and is either already in Heaven as an immortal being or will be in Heaven as an immortal being after its sojourn in corporeality ends; and (5) *a mortal being* is a spiritual being whose eternal soul is fallen and is

(a) in corporeality or (b) in incorporeality either waiting to be born or waiting to be judged.

For the present author, *intellect,* or *intelligence,* denotes the capacity for learning in relationship to reasoning and memory; *intellect,* or *intelligence,* includes cognitive, emotional, physical, social, and spiritual awareness. In subtle contrast to *intellect,* or *intelligence, consciousness* denotes awareness of one's self, or personal being, in relation to one's past surroundings, immediate surroundings, and perceived future surroundings; like *intellect, or intelligence, consciousness* includes cognitive, emotional, physical, social, and spiritual self-awareness.

A newly-understood concept enlarges the consciousness of a person as well as the entire human race by virtue of the person belonging to the human race. Ideally, of course, a newly-understood concept needs to be shared with at least one other person who then can carry its torch further.

In its best application, Aristotle's sound reasoning explains why a tri-unity of the Creator-God is difficult for many people to understand:

> Now there are several ways in which the one and the many are in opposition. One of these lies in the fact that the one and the many are opposed as indivisible and divisible. What is either divided or divisible is accounted for as a kind of plurality, whereas what is indivisible or not divided is said to be a unity.
>
> *Metaphysics, Iota.3, p. 293*

To be sure, in order to understand the tri-unity of the Creator-God, students of Christian metaphysics must hold the whole Godhead while simultaneously attending to the Creator-God's three parts. *Tri-unity* here is not an oxymoron; because the tri-unity of the Creator-God does not oppose itself, the triune aspects of the

Creator-God can be simultaneously and synchronously present. In contrast, Aristotle explained accurately that "it is only opposing things that cannot be simultaneously present" *(Metaphysics, Iota.5, pp. 300-301)*.

Students of Aristotle need to be reminded that, according to Aristotle, *First Philosophy,* or metaphysics, is equivalent to wisdom, theology, and ontology (i.e., the science of being, or the study of "being *qua* being"). Unfortunately, many students of theology have completely separated philosophy from theology, which does the greater disservice to theology. *For example,* to a fault, Christian Science is a religion and not a philosophy because there is only one acceptable view to Christian Scientists on how to interpret the Holy Bible and that view is the view of Mary Baker Eddy. In a philosophy, different views may be compatible and different views may be held by one person at the same time. Eddy's ultimate insult to Evil is to ignore it, ironically, at a cost to absolute truth in promotion of error. Unfortunately for its adherents, Christian Science is able to bury its mistakes, which then can be conveniently forgotten.

Blending Aristotle's *First Philosophy* with Christian metaphysics enables humanity to come to an understanding that the ultimate end point is the saved soul of a human being. Aristotle stated: "And nothing lies beyond an end point. The end point is the extreme in all cases and comprises everything else" *(Metaphysics, Iota.4, p. 296)*. In direct contrast to Plato's and Eddy's thinking, Aristotle also stated that "there cannot be Forms of the kind that some suppose. For then there would be perishable man and imperishable Man" *(Ibid., p. 314)*. Of course, Aristotle's conclusion is inconsistent with Platonism, Neoplatonism, and authentic Christian metaphysics.

Aristotle's views included that metaphysics is the highest form of theoretical science because "in this, if anywhere, would we find divinity" *(Metaphysics, Kappa.7, p. 335)*. Aristotle believed that "there are three kinds of theoretical science: physics, mathematics, and theology" *(Ibid.)*. Because of Hawking's unique view that the laws of science replace the need for God, the present author believes that Hawking would agree with Aristotle's sense of *theoretical science:*

> And the highest kind of science is the theoretical kind, and of theoretical sciences the highest is the last in our list ["theology"]. It has to do with the most valuable of the things that are, and it is the proper object of a science that determines its relative excellence. [brackets mine]
>
> *Metaphysics, Kappa.7, p. 335*

Concerning the physically-knowable universe, anything that is not essential for cosmic evolution, biological evolution, and consciousness evolution as they relate to the salvation of mankind — or even the salvation of one human being — is accidental and co-incidental. "So the infinite must be present as an accidental feature [of perceptibles in the physically-knowable universe]" *(Metaphysics, Kappa.10, p. 343, brackets mine)*. To be sure, all evolution is process and movement across relative time and relative space in the physically-knowable universe. The opposite of such process and movement would be *stasis,* the antithesis of *intelligent evolution* (excluding *homeostasis,* of course).

According to Aristotle, "substance is [the] primary constituent" of the physical universe *(Metaphysics, Lambda.1, p. 355, brackets mine)*. Aristotle's idea of *generated matter* and *ungenerated matter* fit well with Einstein's mathematical construct of mass-energy equivalence as $E = mc^2$ if we assume that Aristotle's *ungenerated matter* represents energy. And Aristotle's idea that "all things originally were in potentiality but not in actuality" *(Ibid.)* is in agreement with what existed within the first three minutes of *the Big Bang* — which is to say, just prior to the earliest nucleosynthesis of the simplest atoms.

The following theological principle that Aristotle articulated may cause the student of Aristotelian thinking to wonder why Aristotle had so much difficulty with Platonism:

> And God also has life; for the activation of thought is life, and He is that activation. His intrinsic activation is supreme, eternal life. Accordingly, we assert that God is a supreme and living being, so that to God belong life and continuous and eternal duration. For that is what God is.
>
> *Metaphysics, Lambda.7, p. 374*

The present author believes that Aristotle would have had great difficulty with the concept of bioevolution because his philosophy required him to conclude that "the primary thing is not the seed but the complete specimen" *(Metaphysics, Lambda.7, p. 375)*. In other words, if asked whether the chicken or the egg came first, Aristotle would have answered that the chicken came before the egg because the chicken was created whole, intact, and sublime. To Aristotle, the hen's egg would not have been thought of as an actual chicken although it would have been thought of as a potential chicken. The present author believes that Aristotle was too focused on matter and its related definitions (which Aristotle was quite good at formulating) in order to be able to grasp or even imagine cosmic evolution, biological evolution, and consciousness evolution in relation to Deity (i.e., the Creator-God). Although Aristotle might have concluded that God is not dead, he probably would have concluded that God just may be asleep. To be sure, we must attribute some of Aristotle's indecision to the unknowable nature of our Creator-God. However, this unknowability only exists when one does not know Christ Jesus or, at least, has no Messianic expectancy.

The present author does not agree with Aristotle that "thinking is the most godlike of things in our experience," but the present author does agree with Aristotle that "absolute thinking is the thinking of thinking" *(Metaphysics, Lambda.9, pp. 382 & 383)*.

The high esteem given to thinking by Aristotle must be counterbalanced with the teaching of Christ Jesus that, for human beings, forgiving love alone approximates divine perfection *(Matthew 5:43-48)*. Thus, unselfish love, and not thinking, is the most godlike of things in our human experience, and love in forgiving others is the beginning of absolute and unselfish love. It is not our intellect that saves us; only the shed blood of Christ Jesus does that.

Applicable to *intelligent evolution,* Aristotle stated: "Traditionally, the evolution of entities must be advanced before either the good or the fine make their entry" *(Metaphysics, Nu.4, p. 445)*. In other words, cosmic evolution, biological evolution, and consciousness evolution had to reach certain levels of advancement before fallen souls could emerge as *Homo sapiens* and, thereby, be in the state and condition appropriate for meaningful salvation opportunities to be offered to them and accepted by them. The concept of *intelligent evolution* (in consciousness evolution) is even demonstrated in the Bible's gradual teaching of: (1) atonement, (2) the remission of sins, and (3) blood sacrifice requirements by the Lord God Almighty.

Finally, concerning Aristotle, not only do individual human beings require time to reach a level of development sufficiently mature enough to grasp the need for personal salvation, the entire human race requires time to reach a level of development sufficiently mature enough for Christ Jesus to return to Earth. (Please be assured that the Creator-God alone determines when Christ Jesus returns.) Unfortunately, "the central doctrine of Aristotle's *Metaphysics"* is "that the foundation of the world is natural substance and not some separate and ideal entity" *(Lawson-Tancred, Hugh* in *Metaphysics, Nu.6, p. 450)*. Indeed, Aristotle did not know that "no one can lay another foundation than that already laid, which is Christ Jesus" *(1 Corinthians 3:11 KJV Paraphrase)*.

Let us now explore the metaphysics of Immanuel Kant as it relates to the concept of *intelligent evolution.*

2.5.2 Insights, Implications, and Applications from Kant

This section has been constructed primarily from the following two literary works of Immanuel Kant: (1) *Prolegomena to any Future Metaphysics* (1783) — referred to as *Prolegomena* in the present author's citations (the plural word *Prolegomena* means "introductory remarks" or "the essentials."); and (2) *The Metaphysical Foundations of Natural Science* (1786) — referred to as *Metaphysical Foundations* in the present author's citations.

In the two books just mentioned, Kant often uses the phrases *a priori* and *a posteriori* relative to certain propositions (i.e., suppositions and judgments). Because those two Latin phrases may not be known or understood by the readers of *Intelligent Evolution,* it is important to explain what they mean as well as their significance to this present work:

2.5.2.1 *a priori* and *a posteriori*

The phrase *a priori* is a Latin prepositional phrase with the preposition "a" meaning *from, out of, based on, after,* or *by way of* and the noun "priori" meaning *the former, the past,* or *the prior.* In common usage, extended meanings of the phrase *a priori* include: *"from the past," "based on one's prior knowledge,"* and *"by way of past individual experience."*

According to Kant, the most substantive types of *a priori* propositions are *synthetical,* or expansive, *a priori* propositions. Here, the phrase *a priori* especially connotes theoretical, speculative, or intuited judgments, assumptions, hypotheses, statements, or ideas that have not yet been tested through additional factual study, analysis, and personal experience (and, in some cases, can never be tested). In other words, although *synthetical a priori* propositions may be assumed based on prior personal knowledge and experience that have produced one's personal worldview and knowledge storehouse, they have not yet been proven by additional knowledge and experience and are, therefore, assumed as true without specific analysis and factual proof. (To Kant, because their truth is self-evident, these *a priori* judgments do not require factual investigation.) Kant would have it that knowledge "lying beyond experience" is knowledge derived *a priori* — or, in other words, "from pure understanding and pure reason" *(Prolegomena, English translation, p. 13).*

As an English speaker (English is my L1, or native language) as well as a German speaker (German is my L2, or second language), the present author finds interesting the difference in the syntactical placement of the phrase *a priori* within Kant's original German and various English translations of his literary works. In German, the prepositional phrase *a priori* is often used after its associated noun. *For example:* "Sie ist also Erkenntnis *a priori,* oder aus reinem Verstande und reiner Vernunft" *(Prolegomena, German original, p. 12).* In the original German, *a priori* functions as a prepositional phrase analogous in syntactical use to the prepositional phrases *aus reinem Verstande* and *[aus] reiner Vernunft.* However, the same sentence is rendered by translators in English as: "It is therefore *a priori* knowledge, coming from pure understanding and pure reason" *(Prolegomena, English translation, p. 13).* The translators would have been more accurate to translate the sentence into English as: "It is therefore knowledge *a priori*, or from pure understanding and pure reason." In the German, "or from pure understanding and pure reason" defines *a priori;* in the English, "coming from pure understanding and pure reason" defines the entire noun phrase "*a priori knowledge*," which particular syntax makes for a slightly different nuanced meaning.

One might call *synthetical,* or expansive, *a priori* propositions *intuited suppositions,* but that description adds another layer of complexity by implying the question "Can accessible *a priori* knowledge actually exist independent of one's personal experience and language acquisition?" The present author's answer to that question is: "No, one must have at least some life experience and a storehouse of some language-related knowledge to posit an untested hypothesis or a conclusion *a priori*." The operative word in the question posed is *accessible*. The present author's answer would have been "Yes" if the question had been "Can *inaccessible a priori* knowledge actually exist independent of one's personal experience and language acquisition?"

Even if a person is susceptible, or sensitive, to receiving external images and ideas through invisible, spiritual, or psychic means, that person must still depend on conclusions made from his or her physical and/or mental past experiences to serve as a filter for mentally testing the authenticity and accuracy of the received images and ideas. It is also legitimate to ask if "past experiences" here might include experiences during one's current corporeal life only or by way of *far memory* — which is to say, *soul memory* from: (1) previous incarnations; (2) past incorporeality *(for example,* during one's existence in between incarnations); or (3) intermittent incorporeality. Here, *intermittent incorporeality includes* out-of-body experiences during astral projection, spiritual visions, supernatural revelations, profoundly deep psychic impressions, and trances[14] induced by the Creator-God's Holy Spirit.

The clearest words that the present author can use to explain *a priori* come from Thomas Jefferson. In the introduction to the *Declaration of Independence,* Jefferson wrote: "We hold these truths to be self-evident." All "self-evident truths" are *a priori* suppositions.

In comparison to the phrase *a priori,* the phrase *a posteriori* is a Latin prepositional phrase with the preposition "a" meaning *from,*

---

[14] **In the King James Version of the New Testament (Acts 10:10, 11:5, & 22:17), the word *trance* is translated from the Greek word ἔκστασις (ekstasis). In other words, a trance induced by the Creator-God's Holy Spirit is a state of spiritual *ecstasy*.**

out of, based on, after, or *by way of* and with the noun "posteriori" meaning *the latter.* Thus, an extended meaning of the phrase *a posteriori* is "based on experimentation" or "after the facts are known." The phrase *a posteriori* connotes propositions, assumptions, hypotheses, statements, or ideas that have been proven, or tested, through factual study, analysis, and personal experience. Therefore, *a posteriori* conclusions have been individually proven, or tested, through specific analyses resulting in factual proofs. While an *a priori* proposition may be someone's mere opinion tested only mentally through reasoned judgment or in one's imagination by way of conclusions made from past experiences, an *a posteriori* proposition is someone's conclusion tested by experimentation, including qualitative analysis and/or quantitative analysis. In other words, conclusions from an *a posterior* proposition rely on empirical data from scientific investigation.

Although some academicians might disagree, understanding where experience fits in is at the heart of cognizing the intended meaning of the phrases *a priori* and *a posteriori: a priori* requires having prior experiences in order to make a hypothesis, and *a posteriori* requires having additional experiences designed to test (i.e., prove or disprove) a specific hypothesis. An example of usage for both phrases is herewith provided by the present author using Chapter One of the Apostle Paul's *Epistle to the Romans.* The following discussion of Chapter One of *Romans* also provides an example of just how essential philosophical discussion is to garnering a more solid understanding of Scripture. (For the sake of clarity, the Holy Bible is the only real Scripture.)

Chapter One of Romans alludes to some people having knowledge *a priori* about the Creator-God based entirely on their observations of His intelligent and miraculous designs in the cosmos and in animate matter (which knowledge is *teleological):*

> {19} That which may be known of God is manifest in these people because God has shown it to them. {20} Because the

> invisible things of God (even His eternal power and Godhead) are clearly seen and understood from the creation of the world through the things that have been made, such people are without excuse: {21} Although they knew God through His creation, they did not glorify Him as God and were not thankful to Him. As a result, they became vain in their imaginations, and their foolish hearts were darkened.
> *Romans 1:19-21 KJV Paraphrase*

In other words, the Apostle Paul believed that people should be able to theorize, infer, or intuit *a priori* that there is a Creator-God based on the existence of what has been created. For these people, viewing what exists as a *creation* presupposes that what exists has been *created* by a *creator* (i.e., a Supreme Being with consciousness) as opposed to viewing it as the product of a series of chance actions and random events. Obviously, in contrast to what the Apostle Paul posited, if one already believes that all physicality and animate matter are results of a series of physical accidents, then one is less likely to theorize, infer, or intuit *a priori* that there is a Creator-God based on what exists because one does not believe that what exists has been created *by* a Supreme Being; instead, one believes that what exists has been "created" — or, rather, *made* — from a series of random physical events. Unfortunately, the Apostle Paul's conclusions do not address people whose worldviews purposely exclude Deity; thus, though not intentionally, the Apostle Paul left an understanding of such skepticism and the basis for a godless worldview to the philosophical explication and psychological discourse of others. In other words, a full discussion is required on how an atheistic belief system influences recognizing, or failing to recognize, the intelligent design of a Creator. Indeed, if people start out already believing that there is no Creator, then nothing short of a supernatural event will convince them otherwise.

In comparison to people who can only theorize, infer, or intuit that a Creator-God exists, authentic Christians are people who know that the Creator-God exists based on their personal experience with His *only-begotten* Son through their own individual salvation and conversion experience and through understanding exactly who Christ Jesus is by reading (or listening to) and comprehending the gospel message in the Holy Bible, especially the New Testament. Thus, authentic Christians have an *a posteriori* knowledge of the Creator-God because they have experienced Him for themselves. (To be sure, Kant and many philosophers would disagree with this conclusion, especially with the way in which the present author has used the phrase *a posteriori*.) Authentic Christians know that the Creator-God exists, and they trust Him because they know that He exists through their personal salvation experience.

Unable to rely on the Holy Bible and a personal conversion experience, Kant could only concede that one may "look beyond this boundary [established by the world of appearances] to the Idea of a Supreme Being" *(Prolegomena, p. 110, brackets mine)* in order to specifically theorize, infer, or intuit a Supreme Being through analogy:

> Thereby [consciousness] does not just *invent* a being, but, as beyond the sensible world there must be something that can be thought only by the pure understanding, determines that something [i.e., the existence of a Supreme Being]... only, of course, by analogy [to the sensible world]. [brackets mine]
>
> *Ibid.*

2.5.2.2 On the meaning of *Science*

Of the two works mentioned at the beginning of the section entitled *Insights, Implications, and Applications from Kant,* the first one listed is the one more heavily relied upon by the present author. The long title of Kant's *Prolegomena to any Future Metaphysics* is actually *Prolegomena to any Future Metaphysics that will be able to emerge as Science* (*Prolegomena zu einer jeden künftigen Metaphysik, die als Wissenschaft wird auftreten können).* Just as *science* was viewed differently from modern science during Aristotle's time, so too was *science* viewed differently from modern science during Kant's time. For that reason, an earlier sense of *science* is herewith provided:

Some insights are gained when one looks up the word *science* in the 1828 edition of Noah Webster's *An American Dictionary of the English Language.* There, the primary purport of *science* is given as "knowledge, or certain knowledge; the comprehension or understanding of truth or facts by the mind." Then, using that sense in example, Webster declares: "The *science* of God must be perfect!" Further, Webster states that the term *science* may be applied to subjects "founded on generally acknowledged truths, [such] as *metaphysics*."

Let us now compare Webster's 1828 etymology of the word *science* with the Hebrew and Greek words from which the word *science* has been translated in the 1611 King James Version (KJV) of the Holy Bible. Webster traced the English word *science* back to the Latin noun *scientia* — which comes from the Latin verb *scio/scire,* originally meaning "to discern or distinguish," only later taking on the sense "to know." Fortunately, for the Hebrew and Greek etymologist, the word *science* is used once in the Old Testament *(Daniel 1:4 KJV)* and once in the New Testament *(1 Timothy 6:20 KJV).* The Hebrew word from which *science* has been rendered is *mad·dä´* [H4093], which means "intelligence" or "consciousness." Stepping to the side and examining that Hebrew word's closely related heteronym *ma·dü´·ah* [H4069], primitive particle *mä* [H4100], and probable root word *yä·dah´* [H3045], we may extrapolate the truer sense of the word *science* in its earliest usage in the English language as "the discovering, discerning, and comprehending of the what, when, why, and how of *being*" — where *being* would logically apply to both inanimate and animate matter and mean "existence." Looking to the Greek New Testament, we find that the word *science* has been translated from *gnō´-sēs* [G1108], a word that has the connotation of "inner knowledge," "knowledge not derived from the physical senses," or "knowledge derived from spiritual or *a priori* cognition." Thus, as with the Hebrew so with the Greek are we brought to an understanding of *science* as "the spiritual sense or intellectual awareness of *being"* — which is in close agreement with the definition for *ontology* as "the science of being."

As a side note, John Wycliffe (d. 1384), the first complete translator of the New Testament into English from St. Jerome's Latin Vulgate, chose the phrase "science of health" (contemporary English spelling is used here) instead of "knowledge of salvation" as found in the King James Version. Thus, Wycliffe rendered the prophecy of the priest Zacharias concerning the Messiah, Christ Jesus, as "he shall bring [the] science of health to his people" *(Luke 1:77 Wycliffe).*

The word *science* did not always mean the systematized knowledge of physicality nor imply a multi-step process for investigation — which is now used in modern science and commonly thought of in terms of the scientific method applied to biology, chemistry, and physics as well as their convergent disciplines and their various extensions in applied areas (i.e., the *applied sciences).* Thus, in its earliest usage, the word *science* conveyed a different meaning than it does today. Previously, it meant "a body of knowledge presupposed to be true *a priori.*"

Paradoxically, there are many Christian fundamentalists today who would object to the use of the word "Science" with "Christ" or "Christian" at the same time that they would feel entirely justified in using the phrase "Creation Science" to describe their posited alternative to the theory of neo-Darwinian evolution. Moreover, those who might object to the nomenclature "Christian Science" would have no problem using the words *theology* and *Christology.* This is especially ironic since the suffix *–ology* means "study, or science, of" and that, thus understood, *theology* may be defined as "the Science of God" and *Christology as* "the Science of His Christ."

2.5.2.3 On the meaning of *Natural Science*

This section is included because the present author believes that many people today who hear or read the words *nature* and *natural* think only of observing wildlife in its immediate environment or in the world at large — in other words, viewing plants and animals in their *natural* habitats. However, *nature* and *natural* includes all inanimate matter as well as all animate matter. Thus, *natural science* includes not only the biological sciences but also the physical sciences and, therefore, any scientific study of the physically-observable universe, or cosmos.

Some modern scientists tend to look at observational studies — like natural history and physical anthropology — as less than academic because all steps of the scientific method might not be immediately employed. However, the recording of pure observations on nature using the naked eye or using technology — like telescopes, light microscopes, transmission electron microscopes, and scanning electron microscopes for cosmological, histological, and geological observations — are part of natural science as *discovery science,* or *discovery-based science*. To be sure, teachers of natural science often use an *inquiry method* to generate interest in their students for the particular natural science they teach, which method begins with questions and hypotheses formulated after making multiple observations.

Before the present author began to read Kant's *The Metaphysical Foundations of Natural Science* (1786), he was hoping to extract some practical understanding about metaphysics and natural science that could be beneficial to twenty-first century learners. Unfortunately, there was little to be found in the book. Perhaps a more accurate title for Kant's *Metaphysical Foundations of Natural Science* might have been *Overarching Theoretical Principles of Motion*. Indeed, *metaphysics* is used in Kant's *Metaphysical Foundations* to represent laws of physics that were incomprehensible, unexplainable, and unknown during Kant's lifetime but are now understood during the twenty-first century. So,

Kant's *metaphysics* in *Metaphysical Foundations* has nothing practical to offer with regard to understanding the *unseen invisible* either in spiritual reality or in *a priori* cognition. Kant's *Metaphysical Foundations* has more to do with phenomenology than noumenology.[15] Even Kant's *transcendentalism* has nothing to do with spiritual reality and everything to do with *transcending,* or rising above, one's current understanding of physics. Kant lived during such a scientifically backward time that he did not even recognize chemistry as a natural science; he stated that "chemistry indeed should be rather termed systematic art than science" *(Metaphysical Foundations, p. 8).* Similarly, Kant believed that psychology was an art and not a science.

Concerning what natural science and metaphysics are, Kant states:

> A rational doctrine of nature deserves the name of natural science only when the natural laws at its foundation are cognized *a priori,* and are not mere laws of experience.
>
> *Metaphysical Foundations, p. 8*

> Pure cognition of the reason from mere *conceptions* is called pure philosophy or metaphysics...
>
> *Metaphysical Foundations, p. 9*

In other words, Kant's *a priori* scientific knowledge has more to do with eureka, gestalt, and epiphany moments regressed, or derived by working backwards, from the sense-world — as well as with intuiting overarching natural laws governing matter — than with recognizing unseen governing principles of the Prime Mover, Deity, or Creator-God.

---

15 ***Noumenology*** **is the study of (1) things-in-themselves, (2) the causes of** ***phenomena,*** **and (3) the nature, or essence, of being;** ***phenomenology*** **is the study of objects of direct experience — which is to say, the physical manifestations of** ***noumena.***

For the present author, the question "Do you know what a cat is?" has subsumed within it the following three separate questions:

1. Do you recognize the physical attributes of a cat?

2. Do you understand a cat's behaviors relative to its personality, its instincts, and how it thinks?

3. Do you understand *a priori* what a cat is in the mind of the Creator-God through His intelligent evolution of it?

For Kant, only the first two questions would have significance. The third question would be superfluous to Kant because it would require supernatural and, therefore, unobtainable knowledge. However, Kant would probably re-ask the third question as "Can we know the meaning of the cat-in-itself *(die Katze an sich selbst)* apart from a cat's physical appearances (i.e., apart from its form)?"

For the student of Christian metaphysics, the following quote from Kant might show promise until the student realizes that, for Kant, *soul* is only the elusive human mind studied by the nebulous "art" of psychology:

> Now Nature, in this sense of the word, has two main divisions in accordance with the main distinction of our sensibility, one of which comprises the objects of the *outer,* the other the object of the *inner* sense; thus rendering possible a two-fold doctrine of Nature: the DOCTRINE OF BODY and the DOCTRINE OF SOUL, the first dealing with *extended,* and the second with *thinking,* Nature [these two doctrines are also called "corporeal doctrine and mental doctrine" on pp. 10-11 of the same work]. [brackets mine]
>
> *Metaphysical Foundations, p. 7*

As viewed by the present author, the primary lesson in Kant's *Metaphysical Foundations* is that without the Creator-God's Holy

Spirit residing in us as a result of our personal relationship with Christ Jesus, *a priori* knowledge and intuitions are merely generalizations, conceptualizations, theorizations, and speculations from the human mind; they are not the knowledge of truth imparted to us by the Creator-God's Holy Spirit. However, in the true spirit of Christian metaphysics, *a priori* knowledge and reason are neither mere postulations nor speculations. To be sure, one cannot really investigate *a priori* sources without personally knowing the Source of all physical things, which Source is also the Source of all things-in-themselves *(die Dinge an sich selbst)* — and which Source is the Creator-God Himself. (The phrase "things themselves" *[die Dinge an sich]* can be used interchangeably with "things-in-themselves" *[die Dinge an sich selbst]).* Although the meaning of *things-in-themselves* is neither supernatural nor esoteric for Kant, the meaning of *things-in-themselves* can be conceptually elusive for many beginning students of Kant. (See the discussion on *noumena* in Section 2.5.2.4 — entitled *The Science of Metaphysics and the Metaphysics of Science.*

In *Metaphysical Foundations,* Kant hinted at the definition of metaphysics, defining it only in Aristotelian terms. In *Assumption Two* of the *Introduction* to *Intelligent Evolution,* the present author stated that Immanuel Kant was an "agnostic and philosopher." Although Kant was an agnostic, he had been immersed culturally in Prussian Pietism. As a result, he had at least been exposed to major Christian concepts and principles about whose certainty he was, or had become, unsure.

In many ways, Kant's *Metaphysical Foundations* is a sketchy regurgitation of Aristotle's *Physics.* To be sure, Kant is pitiable in this book for multiple reasons, including that, in his thinking, there is no room for spiritual insight and supernatural revelations. The theme of his book is also poorly explicated. Kant himself expressed on page 17 of *Metaphysical Foundations* that he did not have sufficient time to devote to its writing. This should be painfully obvious to its readers.

2.5.2.4 The Science of Metaphysics and the Metaphysics of Science

Although Kant's *Metaphysical Foundations of Natural Science* was articulated poorly, Kant's *Prolegomena to Any Future Metaphysics* is eloquent. In his *Metaphysical Foundations,* Kant delivers on phenomenology but not on noumenology. In his *Prolegomena,* Kant delivers on both noumenology and phenomenology.

For the sake of clarity, *noumena* (singular, *noumenon*) are the unseen *things-in-themselves* behind all *phenomena* (singular, *phenomenon*) of physicality. (See also the present author's Footnote 15.) Kant is clear that, although we can recognize physical appearances through our experiences as well as potential experiences with material objects, we can never know their *noumena,* or their truer meaning(s), unless we are contemplating them by using pure reason or regressing (i.e., working backwards) analytically from the phenomena that represent them.

To be sure, Kant is fundamentally Aristotelian in his thinking on metaphysics. In *An Introduction to Volume One,* the present author states that *metaphysics* is "the branch of philosophy and theology that includes the studies of *being* and *reality* (visible reality as well as invisible reality)." For people who focus primarily on the supernatural and esoteric, "invisible reality" might only refer to *spiritual reality.* In contrast, for people who focus primarily on the natural and exoteric, "invisible reality" might only refer to intellectual understanding and reasoning in the mental sphere of generalizations, conceptualizations, theorizations, speculations, and intuitions. Of course, these two views do not need to be mutually exclusive; however, an "either-or" view is maintained by many people relative to: (1) *metaphysics,* (2) how to define *metaphysics,* and (3) how to apply *metaphysics.* Thus, before you enter into a discussion on metaphysics with others, it is best that you know their fundamental views on metaphysics before having the discussion. In other words, do they believe that the invisible reality of metaphysics is only spiritual, only mental, or both spiritual and mental? For Kant, "invisible reality" refers only to pure understanding and pure reason in the mental sphere of generalizations, conceptualizations, theorizations, speculations, and intuitions.

Because Kant is fundamentally Aristotelian in his thinking, what can we find in his *Prolegomena* that might be helpful in building a cohesive understanding of Christian metaphysics as it relates to *science* generally and *intelligent evolution* specifically? And, if Kant's unique vocabulary itself was not intended to possess supernatural and esoteric meanings, what insights, implications, and applications can we gain from the unique language in his *Prolegomena* to apply to *intelligent evolution?*

Kant asked: "If [metaphysics] be science, how is it that it cannot, like other sciences, obtain universal and lasting recognition?" *(Prolegomena, p. 3, brackets mine)* Kant answered his question by stating that, in metaphysics, there is "no standard weight and measure to distinguish sound knowledge from shallow talk" *(Ibid., p. 4)*. Kant mused that if only we could make "the connection of cause and effect (including [their] derivatives [of] force and action)" *(Ibid., p. 6, brackets mine)* as well as other similar connections *a priori,* then there could be "a complete reform of the science [of metaphysics]" *(Ibid., p. 7, brackets mine)*. For the present author, the following fundamental question presents itself: "Can we attain metaphysics as a perfectly new science and way to understand the Creator-God's *intelligent evolution* without making it cultish?" "Yes, we can" is the present author's answer.

In *Prolegomena*, Immanuel Kant contrasted his critical idealism to dogmatic idealism, skeptical idealism, and mystical idealism. As the present author sees it, these three contrasting idealisms default to cults of positivity, immaterialism (i.e., matter as illusion), and mystery religion when they are not grounded in the person of Christ Jesus. In comparison to these three idealisms, Kant called his idealism "critical" because his was an attempt to raise idealism to a scientific level — a level at which it did not exist before Kant (and often does not exist today). In doing so, Kant tried to establish a need for (1) principles, (2) theorems, and (3) steps in metaphysics in order for it to be properly called a *science* — a science not derived from experience but from pure understanding and pure reason. To this end, the present author uses the phrase *scientific metaphysics* synonymously with Kant's *critical metaphysics.*

Is metaphysics for everyone? Although modern science in one form or another, and at one level or another, is for everyone, metaphysics may not be for everyone. Kant stated:

> ...many minds will succeed very well in the exact and even in deep sciences more closely allied to the empirical, while they cannot succeed in investigations dealing exclusively with abstract concepts.
>
> *Prolegomena, p. 11*

Following are the eight major theorems of scientific, or critical, metaphysics that the present author has extrapolated from Kant's *Prolegomena:*

(1) Scientific metaphysical knowledge cannot be empirical.

(2) Scientific metaphysical knowledge is beyond human experience.

(3) Scientific metaphysical knowledge is knowledge *a priori* — which is to say, knowledge from pure understanding and pure reason as well as from intuition, speculation, inference, and imagination.

(4) Scientific metaphysical knowledge uses abstract concepts and articulates them in understandable language. In fact, metaphysical concept elaboration precedes metaphysical practice (i.e., looking at life metaphysically and using metaphysics to help solve some of life's challenges).

(5) Scientific metaphysical knowledge expands human consciousness. (Scientific metaphysics adds something to, or amplifies, concepts human beings already possess.)

(6) Scientific metaphysics is ever-expanding because the entire knowledge base for human beings is ever-expanding.

(7) Although the truths of some scientific metaphysical concepts, propositions, and judgments are self-evident, all scientific metaphysical concepts, propositions, and judgments should be analyzable according to established objective criteria.

(8) The field of scientific metaphysics monitors itself to avoid defaulting to cultish conceptual frameworks of idealism or immaterialism.

Following are seven additional theorems used by the present author to distinguish Christian scientific metaphysics from philosophical scientific metaphysics:

(9) The difference between understanding invisible reality that is mental and intellectual and understanding invisible reality that is spiritual and supernatural is obtained only in the presence of the Creator-God's Holy Spirit.

(10) Without an abiding, authentic faith in the sacrificial atonement of Christ Jesus, practical Christian metaphysics is not possible for individuals.

(11) Although Christian metaphysics has rules, it is not dogmatic except for the role of Christ Jesus in salvation.

(12) Christian metaphysics achieves success solely by never giving up — in being consistently unfailing, unfaltering, and unwavering in devotion to Christ Jesus. In other words, Christian metaphysics is consistently unfailing, unfaltering,

and unwavering regardless of its results when applied to challenges in the world of appearances.

(13) Because objective truth is found in Christ Jesus, Christian metaphysicians are not merely speculative philosophers.

(14) Behind each physical thing, every physical experience, and every potential physical experience, there is at least one associated metaphysical concept.

(15) The reliability, demonstrability, and provability of Christian metaphysics are found in its teachability, practicality, and usefulness.

Critical metaphysics and modern science are similar in that the entire knowledge base for each is ever-expanding. If the knowledge base for either of them ever stagnates, then there can be either no critical metaphysics or no modern science for the individual, the culture, the community, the organization, or the nation-state for which it stagnates.

When metaphysics stagnates, it is no longer scientific metaphysics. And without scientific metaphysics, there can be no true metaphysics (or *critical metaphysics* in the language of Kant). *For example,* without its ability to conceptually expand, some systematic theologies have deteriorated into mere cults of positivity. Without the Creator-God's Holy Spirit as their source of inspiration, these theologies have become stagnant and dying or dead. Locked in religious dogma, their bureaucratic organizations exist only to perpetuate themselves. They have rendered themselves and their adherents incapable of new discoveries. They cannot expand because of the constraints they have placed on themselves. As a result, they are neither scientific nor metaphysical.

When science stagnates, then it is no longer modern science. And without modern science, there can be no true science. *For example,* without an ability to conceptually expand through scientific research, some alternative healing practices have become, or remain, pseudosciences. Locked in pseudo-scientific dogma, their artistry exists mostly to perpetuate their own practices. Because they have rendered themselves and their practitioners incapable of new supportive discoveries through bioscientific, evidence-based, and translational research, they are neither scientific nor metaphysical. They cannot expand because of the constraints they have placed on themselves.

Kant was clear in his belief that we can only know, experience, and intuit based on our sense perceptions of things. For this reason, he stated that "we can know objects as they only *appear* to us (to our senses), and not as they are in themselves" *(Prolegomena, p. 30, Kant's parentheses). For example,* without experiential referents, plane geometry could not exist as a mathematical science. And the proofs required in Euclidean geometry (the high school geometry commonly taught during the twentieth century) using theorems, postulates, axioms, and hypotheses disprove the need for self-evident, or *a priori,* certainty in its mathematical science.

According to Kant, the best we can do is conceptually regress [work backwards] from phenomena to derive *a priori* or intuit the *noumena* behind the phenomena to which we are exposed, grasping at the same time that "[physical] objects are not representations of things as they are in themselves" *(Ibid., p. 33, brackets mine).* If they had collaborated, Aristotle and Plato might have stated the same thing as "forms are not Forms."

In contrast to relying only on their sensory perceptions of physical phenomena, Christians are taught not only to anticipate spiritual phenomena but also to rely on spiritual discernment as they look forward to the future — when they will know to the same extent that they themselves are known by the Creator-God *(1 Corinthians 13:12 KJV).* And, although there are mirror images between the spiritually-observable universe and the physically-observable universe, the mirror images are congruent only in the Mind of the

Creator-God and His Holy Spirit. In other words, the Creator-God is able to hold *the Whole Universe* at the same time that He simultaneously attends to all past, present, and future phenomena in the physically-observable universe as well as to their *noumena* in the spiritually-observable universe.

In order to understand the concepts belonging to the paradigm of *intelligent evolution,* it is worthwhile to reiterate that, according to Kant, "the senses never and in no manner enable us to know things in themselves" *(Prolegomena, p. 36).* Thus, according to the present author, in order to understand fundamental concepts belonging to *intelligent evolution,* we must spiritually theorize, infer, and intuit what was in the Creator-God's Mind from the inception of organic molecules and supramolecular assemblies to the various stages of microevolution and macroevolution throughout the history of the planet Earth. Fortunately, just as kinetoscopic images were available to Gertrude Stein that she might imagine differently from those who lived in generations that preceded her, so too are digital images available to us in the early twenty-first century to help us imagine differently from those who preceded us. Now, through digital imagery and informational graphics, human beings can easily picture how organic molecules and supramolecular assemblies can take shape and be built on invisible templates — such invisible templates, in the case of *intelligent evolution,* provided by the Creator-God Himself as the divine mental fabric upon which the stages of *intelligent evolution* are constructed. In other words, if *abiogenesis,* or the development of organic molecules from inorganic and inanimate substances, occurred within a lightning-charged primordial broth, then it is the Creator-God Himself who made the soup as well as stirred and simmered it until it was done.

Although Kant admits to idealism in the form of *critical idealism,* he seeks to avoid *idealism proper,* which, he states, has a tendency to dismiss all physicality as an illusion:

> Hence we may at once dismiss an easily foreseen but futile objection, "that by admitting the ideality of space and of time the whole sensible world would be turned into mere sham."
>
> *Prolegomena, p. 37*

Applicability of Kant's work on scientific, or critical, metaphysics to the concepts underlying the process of *intelligent evolution* is expressed in the two quotes that follow:

> ...whenever we connect our intuitions of sense (whatever they may contain) in space and in time, according to the rules of the coherence of all knowledge in experience, illusion or truth will arise according as we are negligent or careful.
>
> *Prolegomena, p. 39*

> My doctrine of the ideality of space and of time, therefore, far from reducing the whole sensible world to mere illusion, is the only means of securing the application of one of the most important kinds of knowledge to actual objects and of preventing its being regarded as mere illusion.
>
> *Ibid.*

As the present author has stated previously: (1) Regardless of whether you "believe in" (which is to say, "accept") the theory of evolution, its major strength is found in the unifying concept that it presents to the human mind for understanding the interrelationship of all life forms on Earth (and, perhaps, throughout the physical universe). And (2) regardless of whether you "believe in" (which is to say, "accept") Biblical creationism, its major strength is found in the unifying concept that it presents to the human mind for understanding the basic sequence in the origin of all life forms on Earth.

In order to articulate good Christian metaphysics, a spiritual line of tension must exist between what is known, understood, and comprehended metaphysically with how Christian metaphysics is practiced authentically. How does the present author know this? During his entire life, the present author has walked on this line of tension as if it were a tightrope between what is seen physically and what is seen spiritually. So, too, the concept of *intelligent evolution* requires unification of spirituality in its native sense with physicality in its native sense. In this way, the paradigm of *intelligent evolution* provides for a pure science of nature that is derived from scientific metaphysics (i.e., critical metaphysics) using Christ Jesus as its foundation. In keeping with the language of the Holy Bible, Christ Jesus is eternally and all-at-once the only *deific Force, creative Logos,* and *divine Principle* and the only articulated, expressed, manifested, and spoken *Word of God* — who, as Sovereign Lord, is never to be diminished, deformed, or defamed in thought, word, or deed. To summarize, good Christian metaphysics can only take place in a mind that employs *a priori* principles provided by the Creator-God's Holy Spirit. Seeing *intelligent evolution* metaphysically in Christ Jesus proceeds only from understanding *intelligent evolution* metaphysically in Christ Jesus.

Although Kant would say that unexplainable things-in-themselves (i.e., *noumena*) have no referent in either physical experience or physical appearance, he at least acknowledged that things-in-themselves exist and that "their possibility depends solely on the reference of the understanding to experience" *(Prolegomena, p. 60)*. Here, unexplainable things-in-themselves and their possibility "do not derive from experience, but experience derives from them" *(Ibid.)*. Similarly, the conceptual framework upon which hangs the concept of *intelligent evolution* is not derived from physical referents, but physical referents are derived from it:

> And we indeed, rightly considering objects of sense as mere appearances, confess thereby that they are based upon a thing in itself, though we know not this thing as it is in itself

> but only know its appearances, namely, the way in which our senses are affected by this unknown something.
>
> *Prolegomena, p. 67*

Unfortunately, Kant believed that outside of physicality there can be no meaning because human beings can only base meaning on physical appearances, physical experiences, and potential physical experiences. Thus, Kant disallowed meaning based on spirituality, spiritual appearances, spiritual experiences, potential spiritual experiences, and spiritual phenomena. To be sure, we should disallow fictionalized accounts and occult conjectures of creation, but we should not disallow genuine impartation of spiritual truth and true implantation of spiritual knowledge by the Creator-God's Holy Spirit. Although Kant allowed for insights and intuitions regarding things-in-themselves, he posited that these insights and intuitions are themselves derived from physical appearances, physical experiences, and potential physical experiences. Thus, based on a modified version of Kant's reasoning, we should be able to work backwards (that is, *regress*) from physical appearances and physical experiences to gain insights and have intuitions about the Creator-God's progression of thought concerning the steps and stages of morphogenesis in the *intelligent evolution* of the various species in the domains and kingdoms of living things.

In contrast to Kant, who thought that *"the understanding does not derive its laws* (a priori) *from, but prescribes them to, nature" (Prolegomena, p. 67, Kant's parentheses),* the present author thinks that *"the understanding derives laws* (a priori) *from, as well as prescribes them to, nature."* In other words, for the present author, human beings can create unifying concepts in categories but only after sufficient physical, mental, and spiritual experiences. *Intelligent evolution* is a major "principle on which the understanding [can] be exhaustively investigated, and all the functions, whence its pure concepts arise, [can be] determined exhaustively and precisely" *(Ibid., p. 70, brackets mine)*. Through the grammar of metaphysical thinking, we can regress to the multiple start points from which the various stages and steps of abiogenesis, biogenesis, microevolution, macroevolution, and speciation progressed. Thus, as the present author sees it, concepts of reflection provide for concepts of connection. In this way is the principle of *intelligent evolution* most elegant for conceptualizing physical evolution through spiritual means.

Kant stated:

> Metaphysics has to do not only with concepts of nature, which always find their application in [physical] experience, but also with pure rational [or mental] concepts which never can be given in any possible [physical] experience whatever. Consequently, [metaphysics] deals: (1) with concepts whose objective reality (namely, that they are not chimeras) and (2) with assertions whose truth or falsity cannot [referring to both the concepts and assertions] be discovered or confirmed by any experience [mental or physical]. [brackets mine]
>
> *Prolegomena, p. 75*

Thus, Kant did not extend validity to supernatural implantation and spiritual impartation from the Creator-God's Holy Spirit: (1) without Whom we cannot hear, see, or experience the invisible reality of connectedness that exists in the *intelligent evolution* of all life; (2) without Whom we do not receive spiritual, emotional, mental, physical, and social healings; and 3) without Whom we are unable to receive or operate our spiritual gifts. Concerning spiritual efficacy, the mind of the modern scientist begs the question "Where is the statistical reliability of these so-called spiritual activities?" The answer is between the parameters of null and one hundred per cent based on the multivariate factors that impinge on and influence their processes, procedures, and results.

How is Christian metaphysics objectively possible? Christian metaphysics is a different species of thought that permits the individual to rise above corporeal thinking and transcend experiences based on physicality. Although Kant was stuck on invisible reality as only intellectual or mental and not spiritual, he did derive these three categories of transcendental *Ideas* from his *critical metaphysics:* (1) psychological Ideas, (2) cosmological Ideas, and (3) theological Ideas. For the sake of clarity, it should be reiterated that *transcendental* for Kant is not representative of *transcendentalism* or Neoplatonism but only of higher-order levels of critical thinking.

If Kant had understood that Bible prophecy is the "testimony of Jesus" *(Revelation 19:10 KJV),* he would not have written the following:

> ...the cosmological Ideas of the beginning of the world or of its eternity... cannot be of any service to us for the explanation of any event in the world itself.
>
> *Prolegomena, p. 79*

Concerning such cosmological Ideas, Kant continued:

> And, finally, we must, according to a right maxim of the philosophy of nature, refrain from explaining the design of nature as drawn from the will of a Supreme Being because this would not be natural philosophy but a confession that we have come to the end of it.
>
> *Ibid.*

With regard to the last quote, the present author's response is that, in acknowledging the Will of the Creator-God relative to *intelligent evolution,* we have not come to the end of natural science but to a greater understanding of it. Although Kant subscribed to critical idealism, Kant was ever the realist in touting the possibility of meaning only through physical appearances as well as through physical and mental experiences. Lest anyone misconclude that Kant's *pure reason* has a supernatural edge to it, Kant confirmed that "pure reason does not in its [transcendental] Ideas point to particular objects which lie beyond the field of [human] experience, but only requires completeness of the use of the understanding in the system of experience" *(Prolegomena, p. 80, brackets mine).*

Not all of Kant's thinking belongs to the past. However, his ideas of the impenetrability of matter were disproved by modern nuclear physics and the cosmological origin of matter during *the Big Bang* and in current stellar events. And Kant's understanding of the soul is found wanting: Kant would say that whatever can be said of the soul before death cannot be said of it after death. Indeed, for Kant, "the death of a man is the end of all experience which concerns the soul as an object of experience" *(Prolegomena, p. 83).* That is simply not true because, although the saved fallen soul, when returned to Paradise, is fused in the substance of Spirit to its Creator-God, it experiences unparalleled joy and love as an object of experience through the Creator-God's adoration. For the sake of clarity here, although the Creator-God adores His creation, He does not worship His creation. In contrast, created souls of God adore their Creator as well as worship Him because He is their Creator-God. Created souls are predicates of the Creator-God and not vice versa.

Although Kant was not correct in his understanding that things-in-themselves, or *noumena,* are not related to appearances or experiences, he is correct in his understanding that "to conceive the soul as a simple substance [*for example,* Spirit], on the contrary, means to conceive such an object (the simple) as cannot be presented to the senses" *(Prolegomena, p. 86, brackets mine).* However, to understand that Spirit is not experienced by the physical senses does not mean that Spirit does not exist or that matter does not exist. They both exist but on different planes of existence, or levels of consciousness. Kant almost accedes to this understanding by stating "if natural necessity is referred merely to appearances and freedom merely to things in themselves, no contradiction arises if we at the same time assume or admit both kinds of causality, however difficult or impossible it may be to make the latter kind conceivable" *(Ibid., p. 91)*. Then, Kant captured the essence of *intelligent evolution* when he stated that "the cause, as to its causal act, could not rank under time-determination of its state; that is, it could not be an appearance [in physical phenomena], but would have to be considered a thing in itself, while only its effects would be appearances" *(Ibid., p. 92, brackets mine)*. Indeed, the Creator-God, as the sole — or First and Final — Cause of *intelligent evolution*, is not the appearance in physical phenomena but the foundation of appearance in physical phenomena.

Can Deity be found in Kant's writings? Yes, but only in immanent, pantheistic, and symbolical anthropomorphic forms, an understanding of which that can only be regressed (i.e., reasoned backwards) from physical cause and effect. For Kant, *symbolical anthropomorphism* "concerns language only and not the object itself" *(Prolegomena, p. 106);* this is in contradistinction to *dogmatic anthropomorphism,* which assigns human characteristics to the Creator-God literally and not figuratively. Thus, using language labels from Kant, that the Creator-God *tasted* human pain and suffering through the experiences of Christ Jesus represents *symbolical anthropomorphism* and not *dogmatic anthropomorphism*.

Although Kant referred to Deity's "eternal reason" and "divine nature" *(Prolegomena, p. 92),* Kant only conceptualized Deity mentally and not spiritually because he lacked the Creator-God's Holy Spirit. To Kant, "the thing in itself at its foundation and its causality remain unknown" *(Ibid., p. 93)*. Such limitations to his understanding and reason existed because he did not have a personal relationship with Christ Jesus.

An additional application from Kant's *Prolegomena* to *intelligent evolution* in general and speciation in particular is found in this statement about *subordinate beginnings:* "every beginning of the action of a being from objective causes regarded as determining grounds is always a *first beginning,* though the same action is in the series of appearances only a *subordinate beginning" (Ibid., p. 94).* In other words, when applied to *intelligent evolution,* the emergence of each new species through *speciation* is really a subordinate beginning in a succession of beginnings. Although *noumena,* or things-in-themselves, are behind micro- and macroevolutionary phenomena, whose events occur in relative space-time, "determining causes as things in themselves... do not fall under conditions of time [or space]" *(Ibid., brackets mine).* As understood in Christian metaphysics, the appearance of each new species is "subject to natural necessity" *(Ibid., p. 95).* In other words, according to the present author, the purpose or mission of *intelligent evolution* is in the emergence of: (1) a suitable species as host, or residence, for fallen souls as well as (2) all ecosystems that collectively support the survivability, sustainability, and thrivability of that host.

For the sake of clarity, only *Homo sapiens* provides a suitable host for the fallen eternal soul; for this reason, the members of *Homo sapiens* are at the pinnacle of *intelligent evolution* regardless of any and all speciation in the domains and kingdoms of living things that occurred after the initial emergence of *Homo sapiens. For example,* although some new types of bacteria and viruses emerged after the origin of *Homo sapiens,* they are not at the pinnacle of *intelligent evolution.* Their appearance is inconsequential to *Homo sapiens* except for their impacts on ecosystems in which human beings are found, and except for their potential impacts on end-time events in the appearance of apocalyptic diseases (i.e., plagues and pestilences) sanctioned and dispensed through the Creator-God's Wrath (i.e., His Justified Anger).

Concerning his transcendental Idea known as the *theological Idea,* Kant stated that "it totally breaks with experience and from mere concepts of what constitutes the absolute completeness of a thing in general; and thus, by means of the Idea of a most perfect primal Being, it proceeds to determine the possibility, and therefore the actuality, of all other things" *(Prolegomena, p. 96).* This statement is fully complementary to the concept of *intelligent evolution* because the sequential actualization of the various living things is perfectly dependent on the Creator-God.

For Kant, all intuition is intelligent intuition dependent on sense perception as opposed to supernatural intuition dependent on spiritual discernment, supernatural implantation, and divine impartation. Kant did acknowledge, however, that complete satisfaction cannot be derived from reason:

> Reason through all its concepts and laws of the understanding which are sufficient to it for empirical use, that is, within the sensible world, finds in it no satisfaction because ever-recurring questions deprive us of all hope of their complete solution.
>
> *Prolegomena, p. 102*

Kant believed that only in the knowledge of things-in-themselves, or *noumena,* "can reason hope to satisfy its desire for completeness" *(Prolegomena, p. 102):*

> We must therefore think an immaterial being, a world of understanding, and a Supreme Being (all mere *noumena*), because in them only, as things in themselves, reason finds that completion and satisfaction which it can never hope for in the derivation of appearances from the homogeneous grounds, and because these actually have reference to something distinct from them (and totally heterogeneous), as appearances always presuppose an object in itself and therefore suggest its existence whether we can know more of it or not.
>
> *Prolegomena, p. 103*

Kant's position was that human beings can never know things-in-themselves. The present author's position is that saved human beings, through the Creator-God's Holy Spirit who resides within them, can know things-in-themselves, or *noumena,* through spiritual intuition, supernatural discernment, spiritual phenomena, and revelation (i.e., divine impartation). Kant failed to understand that the Creator-God is objectively real to human beings through Christ Jesus. Kant was an agnostic; in other words, he believed that the Supreme Being is "unknown to us" *(Prolegomena, p. 107)* as well as unknowable by us. For himself, Kant confessed that "the nature of the Supreme Cause itself remains unknown to me" *(Ibid., p. 108)*. Indeed, Kant had no personal relationship with Christ Jesus.

The following two quotes from Kant serve as good summary statements for his understanding of critical, or scientific, metaphysics and the bounds of natural theology:

> The world of sense contains merely appearances, which are not things in themselves, but the understanding, because it recognizes that the objects of experience are mere appearances, must assume that there are things in themselves, namely, *noumena*.
>
> *Prolegomena, p. 109*

> Natural theology is such a concept at the boundary of human reason, being constrained to look beyond this boundary to the Idea of a Supreme Being.
>
> *Prolegomena, p. 110*

In other words, because there is an invisible reality behind what we physically see and experience, let reason: (1) fully enlarge itself up to its boundary (i.e., one's psychic horizon); and (2) permit the consciousness of which it is a part to look beyond that boundary. How do we permit our consciousness to look beyond the boundary of reason? The present author responds: "By letting our imaginations soar in keeping with the Will of the Creator-God through His *only-begotten* Son, Christ Jesus."

The language of mathematical science provides a conceptual framework upon which one can think about abstract mental concepts. Although much verbiage in the language of mathematical science is derived from experience, and although some mathematical formulas can have immediate practical application(s) to the world of appearances, some specific aspects of mathematical science are purely theoretical and, therefore, neither derived directly from experience nor have immediate practical application(s) to the physically-knowable universe. To be sure, some theories based on mathematical science may not be proved for decades, or even centuries, if ever. *For example,* mathematical formulas associated with gravitational waves in Einstein's general law of relativity were purely theoretical when they were conceived by Einstein in 1918 and are only being proved as true nearly a century after their formulation.

Similar to the language of mathematical science, the language of Christian metaphysics provides a conceptual framework upon which one can think about abstract spiritual concepts, including those associated with *intelligent evolution.* Although much verbiage in the language of Christian metaphysics is derived from the thinking and experiences of early metaphysicians, and although some Christian metaphysical formulas can have immediate practical application(s) to the world of appearances, some specific aspects of Christian metaphysics are purely theoretical and, therefore, neither derived (i.e., *regressed)* from experience nor have immediate practical application(s) to the physically-knowable universe. To be sure, the theories of *intelligent evolution* may not be proved for decades, if ever, to the satisfaction of researchers. However, as they relate to *intelligent evolution,* the theorems, postulates, axioms, and hypotheses of Christian metaphysics must be clearly stated and explicated if *intelligent evolution* is ever to be tested and proved by additional research in the field of Christian metaphysics and its various domains.

Understanding the existence of language in human beings includes cognizing at least these two major principles: (1) Human language consists of intelligible utterances and/or gesticulations with acknowledged rules for word meaning, grammar, and syntax. (2) The origin of rules for word meaning, grammar, and syntax come from (a) creativity in our cerebralization of human experience coupled with (b) our far memory of the language we once used in immortality before the Adamic Fall.

The brain of *Homo sapiens* has the capacity not only to understand language but also to invent language in its absence, including a language for Christian metaphysics. In learning a second language, or L2, there comes a point in the learning of it when the words, phrases, and sentences of one's native language, or L1, intersect the words, phrases, and sentences of one's L2 at coincident points of meaning in the brain. After one has tried to think in one's L2 long enough, words, phrases, and sentences from one's L2 mentally appear at the appropriate time and place to permit one to think as well as express oneself in that language. Similarly, when one has tried to think in the language of Christian metaphysics long enough, words, phrases, and sentences that convey metaphysical meaning appear in one's thought processes at the appropriate time and place. (To be sure, readiness of the individual as well as need determine the *appropriate time and place*.)

There exists a world of appearances in addition to the one with which Kant was acquainted. Kant was only acquainted with a world of appearances associated with the physically-observable universe. There is also a world of appearances associated with the spiritually- or metaphysically-observable universe. The Apostle John wrote that some elements of that spiritual world would be obvious to people when Christ Jesus returned for his millennial reign on Earth:

> {1} Behold, what manner of love the Father has bestowed upon us, that we should be called the sons [or *heirs*] of God: Therefore, the world knows us not because it knew him not. {2} Beloved, now are we the sons [or *heirs*] of God, and it does not yet appear what we shall look like, but we know that, when Jesus Christ shall appear, we shall look like him

> for we shall see him as he is. {3} And every person that has this hope purifies himself or herself, even as Jesus Christ is pure. [brackets mine]
>
> *1 John 3:1-2 KJV Paraphrase*

In other words, when Christ Jesus appears, each saved fallen soul shall receive its new, personal somatic identity, which is a spiritual appearance (i.e., the *astral gelatinous™* form described on pages I-13 and I-14) and not a physical appearance. Then, we shall see in Spirit even as we are seen in Spirit. And we shall know even as we also have been known, and always will be known, in Spirit. The present author's point here is that not all appearances are physical; some appearances are spiritual. For the sake of clarification, spiritual appearances are *noumena* and not *phenomena*.

Kant stated that if we use metaphysics "as a natural disposition of reason" *(Prolegomena, p. 114),* then metaphysics is actual and can be scientific. However, if we use metaphysics to debate the existence of illusions, then its pursuit is in vain and cannot be scientific. True science includes chemistry and astronomy. Pseudoscience includes alchemy and astrology. Whether metaphysics is science or pseudoscience depends on how it is defined and used by its practitioner(s).

Kant stated that "heretofore [before his time] metaphysics has never existed as a science" *(Ibid., p. 117, brackets mine).* However, Kant continued, if we ground metaphysics in critique and *a priori* propositions and not probability and conjecture, then and only then can it exist as science. The present author adds that metaphysics must not only be grounded in pure reason but also in an understanding of God's written word, the Holy Bible. In contrast to Kant, the present author believes that the authentic Christian does not need to forego metaphysics and its instruction in order to adopt a rational faith. As in the case of *intelligent evolution,* metaphysics and a rational faith can be blended together to see just what the Creator-God has done, and why He has done it.

Let us now explore the metaphysics of Mary Baker Eddy as it relates to the concept of *intelligent evolution*.

2.5.3 Insights, Implications, and Applications from Eddy

Mary Baker Eddy (1821-1910) was the Discoverer and Founder of Christian Science. Christian Science is a dogmatic, cultish religion that combines Neoplatonism and immaterialism. (*Neoplatonism* is a resurgence of Platonism with diverse reinterpretations and extraplatonic inclusions. *Immaterialism* is the belief that physical things have no reality apart from one's perception.) Christian Science pits spirituality against corporeality instead of Good against Evil. It does not just present that corporeality is *delusory* (which it can be) but that corporeality is *illusory* (which it is not). Unfortunately, it misdirects people to fight against the ills of corporeality and not against the demonic forces of Evil. Consequently, time, effort, and energy are wasted by people who try to force change on corporeality even when it is clear that the desired change will not occur. This last statement is not meant to diminish the authenticity of miraculous healings that have taken place throughout Christendom because of divine intervention through prayer, the laying on of hands, and affirmations of the Creator-God's mercy and goodness in the name of Christ Jesus.

To be sure, the ills of corporeality and the demonic forces of Evil do not always overlap: It is Evil, not corporeality, that generates illusions. Corporeality is not an illusion and, except for magic, hypnotism, and certain types of propaganda, it cannot be used to fabricate illusions. And, despite what Eddy has written, corporeality is not an illusion of Evil. Although Evil is not an illusion, it can, and does, fabricate illusions. Evil fabricates illusions in the human mind by using one's: (1) unpleasant memories, (2) unholy desires, (3) unhealthy emotions, and (4) faulty rationalizations:

(1) "Unpleasant memories" here include the memories of past sins, especially unconfessed sins. (2) "Unholy desires" here include lust, greed, covetousness, and vengefulness. (3) "Unhealthy emotions" here include hatred, unforgiveness, jealousy, discouragement, and naiveté about, or indifference to, Evil. And (4) "faulty rationalizations" include making excuses for one's own unholy desires, unhealthy emotions, and sin.

Evil can also use unmitigated pain, depression, suffering, and tribulation to distract us, wear us down, and wear us out in order to more easily implant its illusory seeds and false scenarios within our minds. Based on the memories, desires, emotions, and conditions just mentioned, demonic forces fabricate illusions within our subconscious/unconscious mind in the hope that we will act on them as if they constitute reality. Paradoxically, as we act on Evil's illusions, they actually become our reality. And the more that we indulge demonic illusions, the more the illusions become entrenched within our conscious functioning self as reality. To be sure, demonic forces desire our descent into their hell. It does not matter to them if our descent is rapid or gradual as long as we are spiraling downward.

Christian metaphysics is not equivalent to *Christian Science*. Christian Science is a religion. Christian metaphysics is a philosophical way of life centered on Christ Jesus. Christian Science has some Christian metaphysics in it, but Christian Science is neither the center nor the circumference of Christian metaphysics. Christian Science demonstrates inflexibility in thinking and in its approach to resolving and solving life's problems, but the idealism of Christian metaphysics demonstrates flexibility in thinking and in its approach to resolving and solving life's problems. Christian Science is inflexible, but Christian metaphysics extols the virtues of thinking theologically, spiritually, philosophically, and judiciously all at the same time. (It is in these ways that Christian metaphysics demonstrates flexibility.) Christian metaphysics provides a healthy, circumspect way to think; Christian Science, however, can make and keep one less than healthy spiritually, emotionally, mentally, physically, and socially. The religion of Christian Science shares some, but not all, characteristics associated with the "Prosperity" Movement, the "Confess It and Possess It" Movement, the "Word of Faith" Movement, and the "Speaking Things Into Existence" Movement. (Please review Section 2.2 in this book — entitled *What Thinking Metaphysically Is Not.*)

Eddy's concept of Christian metaphysics as Christian "Science" stems not only from the etymology of the word *Science* (please review Section 2.5.2.2 in this book — entitled *On the Meaning of Science*) but also from the: (1) demonstrability, (2) reproducibility, (3) teachability, (4) practicability, and (5) provability of metaphysical healing. Indeed, as Eddy stated, "Man is deathless, spiritual" *(Science and Health, 266:29),* but, as the present author would add, it is only immortal man restored by the shed blood of Christ Jesus that is deathless and spiritual: It is only saved fallen man, not mortal man, that "coexists with God and the universe" *(Ibid., 266:31-32).*

Christian Science is dogmatic because it does not acknowledge the multivariate nature of corporeality, specifically that there can be multiple contributing factors (sometimes synchronous, sometimes sequential) to the individual ills and negative circumstances of humankind. And Christian Science is dogmatic because it constrains its followers to a specific spoken and written vocabulary and a narrow way of looking at life and dealing with life's problems. For these reasons, Christian Science is not *scientific* metaphysics, which is both self-critical as well as expansive. Because Christian Science cannot expand, it stifles spiritual growth. And, because it does not breathe, it only permits shallow breathing in its followers. Many of its followers have fooled themselves into believing that they are practicing Christian metaphysicians if they wear a smile, ignore life-threatening conditions, and speak positively concerning all aspects of life, including sin, sickness, disease, disability, and death.

Nevertheless, despite all of the negative things that the present author has just written, much can be learned about Christian metaphysics from the writings of Eddy. Eddy was a superior thinker with superior literary skills who established clear connections between Christianity and historical metaphysics in a well-thought-out systematic theology. Indeed, she was the first person, male or female, to establish a systematic theology based on Christian metaphysics. Although other people after her purloined ideas from her to begin their own religious movements, their brands never measured up to her brand: Their brands were only watered down versions of hers because they lacked substance, commitment, and action based on informed faith. Paradoxically, however, remaining steadfast to her singular perspective was the downfall of Eddy's Christian Science. The most unfortunate thing about the Christian metaphysics of Eddy is that it leads people to depend mostly on her written works rather than diligently search the Holy Bible and learn its truths for themselves. As a result, adherents to her theories are often spiritually unbalanced and unhinged from mainstream Christian thinking that is Biblically authentic and theologically sound.

Eddy's greatest fault in the development of her systematic theology was in failing to declare unequivocally the necessity for the sacrificial atonement of Christ Jesus for the forgiveness of sin. Her Christian metaphysics recognized the restoration of *the Whole Universe* (using the present author's phraseology) only to spiritual sense through spiritual *unfoldment*. As previously stated in this book, *unfoldment* is the gradual understanding of the truths in the spiritually- or metaphysically-observable universe and their practical applications to the human experience. *Unfoldment* does not include the restoration of fallen, mortal souls to immortality because Eddy's brand of Christian metaphysics did not posit unambiguously that the truth of all being is found only, and alone, in the shed blood of Christ Jesus. Eddy did not subscribe to the theological position that *mortal man* is *fallen man*. Rather, Eddy posited that *fallen man* is an illusion to the corporeal senses, themselves the source of all illusion and error.

Eddy's answer to the question "Is there no sacrificial atonement?" *(No and Yes, 33:12)* misses the mark. (The present author has often thought that Eddy's just-cited work on this topic might as well have been called *Maybe* instead of *No and Yes.*) Nowhere in any of her literary works does Eddy explicitly refer to the forgiveness of sin as a consequence of the shed blood of Christ Jesus. It should be noted, however, that Eddy did write that "the spiritual essence of blood is sacrifice" *(Science and Health, 25:3)*. Had she just clarified the role of Christ Jesus as the Creator-God's *only-begotten* Son and his death as the only sacrifice acceptable to God the Father for the forgiveness of sin, the present author would have been satisfied with Eddy's treatment of the topic. Without this clarification, Eddy is definitely not on point. Without the shedding of Christ Jesus' blood, there can be no "dominion over all the earth and its hosts" *(Ibid., 102:14-15)*.

Scripture clearly states that *blood is life* and that *there is no remission of sins without the shedding of blood:*

> For the life of the flesh is in the blood: and I [the Lord God Almighty] have given it to you upon the altar to make an atonement for your souls: for it is the blood [of sacrificed animals] that makes an atonement for the soul. [brackets mine]
>
> *Leviticus 17:11 KJV Paraphrase*

> ...the blood is the life.
>
> *Deuteronomy 12:23a KJV Paraphrase*

> And almost all things are by the Law of Moses purged with blood; and, without the shedding of blood, there is no remission of sins.
>
> *Hebrews 9:22 KJV Paraphrase*

During Old Testament times, it was the shed blood of sacrificed animals that regularly made atonement for the sins of Israel.

During New Testament times, it is the shed blood of Christ Jesus that makes atonement for the sins of humankind "once for all" *(Hebrews 10:10 KJV)*. The Lord God Almighty Himself provided a sacrifice for the sins of the world in His *only-begotten* Son, Christ Jesus. Therefore, unless we personally accept the shed blood of Christ Jesus as the only sacrifice acceptable to the Lord God Almighty for the remission of our individual and collective sins: (1) we have no righteousness in His sight, (2) we are not saved, (3) we cannot go to Heaven, and (4) we are subject to His Wrath both on Earth and in the hereafter. (For the sake of clarity, the *Wrath* of the Creator-God is His *Justified Anger.*)

In answering the question "Is there no sacrificial atonement?" Eddy stated:

> The real atonement – so infinitely beyond the heathen conception that God requires human blood to propitiate His justice and bring His mercy – needs to be understood.
>
> *No and Yes, 34:19-22*

> It was not to appease the wrath of God, but to show the allness of Love and the nothingness of hate, sin, and death, that Jesus suffered.
>
> *Ibid., 34:11-13*

> He [Christ Jesus] atoned for the terrible unreality of a supposed existence apart from God.
>
> *Ibid., 34:15-16*

> The spiritual interpretation of the vicarious atonement of Jesus, in Christian Science, unfolds the full-orbed glory of that event; but to regard this wonder of glory, this most marvellous demonstration, as a personal and material bloodgiving — or as a proof that sin is known to the divine Mind, and that what is unlike God demands His continual presence, knowledge, and power, to meet and master it — would make the atonement to be less than the AT-ONE-

> MENT, whereby the work of Jesus would lose its efficacy and lack the "signs following."
>
> *Ibid., 34:11-20*

In trying to address the metaphysical aspects of Christ Jesus' sacrificial atonement, Eddy loses the heart of the matter — albeit in articulately flowing and beautifully wordsmithed language. Unfortunately, because her language is too figurative in addressing the question of sacrificial atonement, it loses the essence of true Christian metaphysics — which enables one to see the visible and the invisible at the same time, mortality and immortality at the same time, Good and Evil at the same time, sin and forgiveness at the same time, death and life at the same time, redemption and damnation at the same time, Heaven and Hades at the same time, and creationism and evolution at the same time. Eddy's figurative use of *atonement* as *at-one-ment* diminishes the significance of the remission of our sins through the blood sacrifice of Christ Jesus. Our *at-one-ment* with the Creator-God is the direct result of sacrificial *atonement,* specifically through the tortured murder and shed blood of Christ Jesus, the *only-begotten* Son of God.

Earlier in this book, the present author wrote that Christian metaphysics is not dogmatic except for the role of Christ Jesus in salvation (Theorem Number Eleven in Section 2.5.2 — entitled *Insights, Implications, and Applications from Kant).* Christian metaphysics must always be dogmatic and precise, however figurative one might be about the role of the shed blood of Christ Jesus in the salvation of humankind. (Remember, just because language is metaphorical does not make it metaphysical and, conversely, just because language is metaphysical does not make it metaphorical.) About the unique and necessary role of Christ Jesus concerning salvation, Christians must be uncompromisingly steadfast and unwaveringly inflexible. Without the sacrificial atonement of Christ Jesus, Christian metaphysics has no power to change anything. In fact, it ceases to be Christian.

In keeping with Eddy, sin *does not* exist inside the Mind of the Creator-God, but, in contradiction to Eddy, sin *does* exist outside of the Mind of God, and the Mind of God identifies sin for what it is. Unfortunately, it did not dawn on Eddy that by naming seemingly illusory things that are opposite and opposing to the immortal life one has in the Creator-God (such nomenclature including *mortal mind, mortal man, disease,* and *death),* she actually acknowledged their existence: One does not name things that do not exist except in fiction. And one does not expend time, effort, and energy to dispel things that do not exist. *For example,* demons cannot be cast out in prayer if one believes that they are mere illusions of Evil. Demons, unclean spirits, devils, or evil spirits (all four terms are synonymous here) have reality and power but, fortunately, not the *ultimate* reality or power. (As a side note, unclean spirits often have a muddy, murky, or brownish cast.)

At the time of this writing, the organized religion known as "Christian Science" is almost extinct because it never established clear-cut theological connections with mainstream Christianity. For that reason, with the death of its leader, organizational Christian Science not only lost its Discoverer and Founder but also its most effective proponent, best apologist, and greatest spokesperson.

Because the overwhelming majority of the earliest converts to Christian Science were from mainstream Christianity, most of them were already knowledgeable about the efficacy of the shed blood of God's *only-begotten* Son, Christ Jesus. Consequently, for them, Christian Science was their *next step* in understanding the Creator-God and practicing the application of His truths to daily living. At the time of this writing, however, virtually no Christian Scientist recognizes or acknowledges the underpinnings of their faith in the shed blood of Christ Jesus. The truth be told (and it is being told right now), the few Christian Scientists who exist at the time of this writing would find God's requirement for an atoning blood sacrifice astonishingly barbaric and uninformed metaphysically. Therefore, without acknowledging the full power of the shed blood of Christ Jesus to appease God's Wrath as well as to save from sin and to heal, Christian Science has lost its greatest power to heal. (Beneficial aspects to positive thinking still exist in it even though it is not perfectly aligned theologically with — which is to say, on the right side of — the Creator-God's absolute truth.) As the present author has stated in his work entitled *As I See It: The Nature of Reality by God (p. 9),* "without the shed blood of Christ Jesus, all spiritual truths are of null effect within our personal lives. To be sure, the truths are not untrue and are not of null effect within the spiritually- or metaphysically-observable universe; they are just 'untrue' in our personal lives — that is, there is no efficacy to their application within our day-to-day experience."

Although the present author recommends that the contemporary student of Christian metaphysics read the copious literary works of Eddy, never mistake that he is an advocate for the institutional bureaucracy known as "Christian Science" (also known as the "Church of Christ, Scientist"). Although some aspects of the doctrines it represents are extremely beneficial, this organized religion has neither grown with the times nor matured into what it might have become had it not backed itself into a theological corner as a self-proclaimed complete, perfect, and intact systematic theology with a "forever Leader." Personally, the present author believes that Christian Science should never have become a Christian denomination; it would have fared much better had it functioned in perpetuity as: (1) an evolving and expanding inter-denominational Christian metaphysical society, (2) an international Christian metaphysics publishing company, and (3) a world class college or university of Christian metaphysics with an undergraduate curriculum similar to that presented in Section 2.4 of this book — entitled *Proposed Curriculum for the Millennium.*

For the present author, the most practical definition of *metaphysics* given by Eddy is in terms of what metaphysics does. In her primary work, *Science and Health with Key to the Scriptures,* she stated that "Metaphysics resolves things into thoughts" *(Science and*

Health, 269:14-15).[16] For the present author, metaphysics not only resolves *things into thoughts* but also *thoughts into things,* which "things" are not only discernible to spiritual sense but are also capable of being apprehended by human beings whose intellect has been properly nurtured physically, emotionally, mentally, spiritually, and socially.

Eddy had a lifelong interest in studying reality. She wrote that as a girl her "favorite studies were natural philosophy [i.e., the systematic study of nature, or *natural science*], logic, and moral science [i.e., philosophy]" *(Retrospection and Introspection, 10:7-8, brackets mine)*. As Eddy matured, her interests expanded to include historical metaphysics. It is evident from the literary works of Eddy that she had a substantial knowledge of well-known philosophers and thinkers who had written about metaphysics, including: Plato (c. 428-347 BC), Aristotle (384-322 BC), René Descartes (1596-1650), Baruch Spinoza (1632-1677), John Locke (1632-1704), Gottfried Wilhelm Leibniz (1646-1716), George Berkeley (1685-1753), David Hume (1711-1776), Immanuel Kant (1724-1804), Johann Gottlieb Fichte (1762-1814), and Georg Wilhelm Frederik Hegel (1770-1831). Indeed, Eddy referred to all of these notable authors in her own literary works.

Although Eddy stated that she had not read Berkeley before the publication of her first edition of *Science and Health with Key to the Scriptures* in 1875 *(Message 1901, 24:21-23)*, by 1901 she wrote the following about Berkeley's work:

> Bishop Berkeley published a book in 1710 entitled "Treatise Concerning the Principle of Human Knowledge." Its object was to deny, on received principles of philosophy, the reality of an external material world. In later publications he declared physical substance to be "only the constant relation

---

16 *Science and Health with Key to the Scriptures* **by Mary Baker Eddy, published by the Christian Science Board of Directors. Boston, Massachusetts, 1906.**

> between phenomena connected by association and conjoined by the operations of the universal mind, nature being nothing more than conscious experience. Matter apart from conscious mind is an impossible and unreal concept." He denies the existence of matter, and argues that matter is not WITHOUT the mind, but within it, and that that which is generally called matter is only an impression produced by divine power on the mind by means of invariable rules styled the laws of nature. Here he makes God the cause of all the ills of mortals and the casualties of earth.
>
> *Message of 1901, 23:23-24:8*

To be sure, Eddy made it clear that she was well-acquainted with earlier works on metaphysics by others when she wrote:

> Leibnitz *[sic],* Descartes, Fichte, Hegel, Spinoza, Bishop Berkeley, were once clothed with a "brief authority;" but Berkeley ended his metaphysical theory with a treatise on the healing properties of tar-water, and Hegel was an inveterate snuff-taker. The circumlocution and cold categories of Kant fail to improve the conditions of mortals, morally, spiritually, or physically. Such miscalled metaphysical systems are reeds shaken by the wind. Compared with the inspired wisdom and infinite meaning of the Word of Truth, they are as moonbeams to the sun, or as Stygian night to the kindling dawn. [brackets mine]
>
> *No and Yes, 22:4-14*

Although Eddy was not entirely correct, Eddy was not all wrong. Indeed, Eddy was a modern-day Hypatia. Both Hypatia and Eddy were vilified for being women with superior abilities in thinking and in articulating their views. Hypatia was a mathematician, astronomer, and philosopher and head of the Neoplatonic school at Alexandria, Egypt during the fourth century of the Christian Era. (Alexandria was the capital of Hellenistic, Roman, and Byzantine

Egypt.) Murdered in circa 405 AD by her political rivals, none of Hypatia's written works survived. Fortunately, the overwhelming majority of Eddy's written works have survived. (Because of their unusual roles in history, it would not surprise the present author to learn, upon his entering Heaven, that Hypatia was a previous incarnation of Eddy.)

Eddy further demonstrated her familiarity with the works of well-known authors of metaphysics by facilely incorporating the vocabulary of historical metaphysics into her writings, which incorporation included the following words and phrases: *a priori, being, corporeality, essence, First Cause, idealism, illusion, incorporeality, Logos, Neoplatonic, noumenon, nothingness, ontology, phenomenon/phenomena, physicality, Platonic, Principle, realism, reality, self-evident proposition(s), somethingness, substance, teleology, transcendental, transcendentalism, universal being, universal law, universal mind,* and *unreality.*

Although it is clear that Eddy used the vocabulary of metaphysics generated by the historical philosophers known to Eddy, it is also clear that Eddy's works were no less original than any other author who has been influenced by their own teachers and mentors as well as significant authors whose works they have read and studied. Indeed, Eddy did not think or write in a vacuum. Although there have been charges that Eddy plagiarized her work from other sources, such charges are completely unfounded. Her systematic theology is original despite its being influenced by people who lived during her lifetime as well as by those who had died well before her time. Concerning the originality of her work, Eddy wrote:

> The first edition of my most important work, *Science and Health,* containing the complete statement of Christian Science, — the term employed by me to express the divine, or spiritual, Science of Mind-healing, was published in 1875. When it was first printed, the critics took pleasure in saying, "This book is indeed wholly original, but it will never be read." The first edition numbered one thousand copies. In

> September, 1891, it had reached sixty-two editions. Those who formerly sneered at it, as foolish and eccentric, now declare Bishop Berkeley, David Hume [who wrote extensively on *moral science*], Ralph Waldo Emerson, or certain German philosophers, to have been the originators of the Science of Mind-healing as therein stated. [brackets mine]
>
> *Retrospection and Introspection, 37:1-15*

> Kant, Locke, Berkeley, Tyndall, and Spencer afford little aid in understanding divine metaphysics or its therapeutics.
>
> *Miscellany, 349:9-11*

Although Eddy distanced herself from one of her most important contemporary mentors and colleagues, Phineas Parkhurst Quimby (1802-1866), and after his death even described him as "an obscure, uneducated man" *(Miscellany, 305:1),* it becomes clear to the discerning and probing student of truth — after examining written accounts of their professional relationship and Quimby's own literary efforts — that Quimby influenced Eddy with regard to (1) her attitude and approach concerning spiritual healing and (2) some of her distinct verbiage. After carefully examining the written evidence, the present author has concluded that, although Eddy's work is original, Eddy brought some of Quimby's ideas to spiritual maturity, fruition, and erudition. To be sure, Quimby had a significant impact on Eddy's thinking.

Throughout her writings, Eddy used these terms and phrases interchangeably: "divine metaphysics," "Christian Science,"[17] "absolute Science," "All Science," "Divine Science," "Spiritual Science," "Mind-science," "the Science of Life," "the Science of Mind," "the Science of Soul," "the Science of Spirit," "the Science

[17] The noun phrase *Christian Science* was published first by Phineas Parkhurst Quimby in 1863. From *The Quimby Manuscripts,* Thomas Y. Crowell Company, 1921, page 388 (see also pages 185 and 196).

of God," "the Science of man," "the Science of Truth," "metaphysical Science," and "the Science of being." Concerning "the Science of being" (which is the definition for *ontology* given previously by the present author), Eddy crafted a "scientific statement of being" from her point of view:

> There is no life, truth, intelligence, nor substance in matter. All is infinite Mind and its infinite manifestation, for God is All-in-all. Spirit is immortal Truth; matter is mortal error. Spirit is the real and eternal; matter is the unreal and temporal. Spirit is God, and man is His image and likeness. Therefore man is not material; he is spiritual.
>
> *Science and Health, 468:9-15*

In contrast to Eddy, the present author has posited that *immortals* are not material in body, or somatic identity, but *mortals* are. Eddy — like Aristotle and Kant — failed to hold *the Whole Universe* while simultaneously attending to its two major parts. As the present author has already explained, the physically-observable universe and the spiritually- or metaphysically-observable universe are the two major parts of *the Whole Universe.*

Concerning the topic of *intelligent evolution,* there is much food for thought in Eddy's writings. Although she herself rejected Darwinism (to be sure, the ever-evolving tenets of neo-Darwinism had not been laid down during her lifetime), Eddy's metaphysical works contained many seeds to help the present author elaborate the concept of *intelligent evolution* (such transcendent seeds unknown to Eddy during her lifetime, of course).

In a paragraph with the margin heading of *Man springs from Mind,* Eddy wrote the following about Darwin and Darwinian evolution:

> All error proceeds from the evidence before the material senses. If man is material and originates in an egg, who shall say that he is not primarily dust? May not Darwin be right in

> thinking that apehood preceded mortal manhood? Minerals and vegetables are found, according to divine Science, to be the creations of erroneous thought, not of matter. Did man, whom God created with a word, originate in an egg? When Spirit made all, did it leave aught for matter to create? Ideas of Truth alone are reflected in the myriad manifestations of Life, and thus it is seen that man springs solely from Mind. The belief that matter supports life would make Life, or God, mortal.
>
> *Science and Health, 543:17-30*

Of course, Eddy's question "May not Darwin be right in thinking that apehood preceded mortal manhood?" is purposely facetious, hoping the reader will make the opposite conclusion. To the Christian Scientist, the last sentence in Eddy's just-quoted paragraph poses no problem because it is assumed that all life is immortal and that no real life is mortal. Eddy did not acknowledge the possibility of two realities each with its own form of *life:* (1) one corporeal and visible and (2) the other incorporeal and invisible. Like Eddy, the present author acknowledges that neither man nor matter constitute the Creator-God, but, unlike Eddy, the present author also acknowledges that the Creator-God has used matter to house in physicality some of His myriad ideas for the sole purpose of providing a complete and perfect cosmological and ecological backdrop to sustain His invention of housing fallen souls in *Homines sapientes* (the plural of *Homo sapiens*) for the purpose of providing them (the fallen souls) with opportunities for salvation.

In a paragraph with the margin heading of *The ascent of species*, Eddy wrote:

> One distinguished naturalist argues that mortals spring from eggs and in races. Mr. Darwin admits this, but he adds that mankind has ascended through all the lower grades of existence. Evolution describes the gradations of human belief, but it does not acknowledge the method of divine

> Mind, nor see that material methods are impossible in divine Science and that all Science is of God, not of man.
>
> *Science and Health, 551:9-16*

In response to the just-cited paragraph, the present author agrees that Darwinian evolution "does not acknowledge *the method* of divine Mind," but the present author posits that the paradigm of *intelligent evolution* does acknowledge that method. Accordingly, *intelligent evolution* is the particular method that the Creator-God used in forming the entire corporeal backdrop that He created *ex nihilo* as summarized in the Genesis account of creation.

Eddy dismissed the contributions of the well-known metaphysical philosophers as well as those of Darwin in understanding the true nature of Man (capitalized here to distinguish from *mortal man,* or humankind):

> When every form and mode of evil disappear to human thought, and mollusk and radiate are spiritual concepts testifying to one creator, — then, earth is full of His glory, and Christian Science has overshadowed all human philosophy, and being is understood in startling contradiction of human hypotheses; and Socrates, Plato, Kant, Locke, Berkeley, Tyndall, Darwin, and Spencer sit at the feet of Jesus.
>
> *Miscellaneous Writings, 361:9-16*

> In its history of mortality, Darwin's theory of evolution from a material basis is more consistent than most theories. Briefly, this is Darwin's theory, — that Mind produces its opposite, matter, and endues matter with power to recreate the universe, including man. Material evolution implies that the great First Cause must become material, and afterwards must either return to Mind or go down into dust and nothingness.
>
> *Science and Health, 547:15-23*

Unfortunately, Eddy was too dismissive of Darwin's understanding of speciation, and her brief explanation of Darwinism is inaccurate and, therefore, unjust.

Although Eddy rejected the notions of Charles Darwin (1809-1882), she did not reject those of Louis Agassiz (1807-1873). To Agassiz, "each species of plant or animal life was a thought of the Creator. This belief was the basis for Agassiz' never-ending opposition to Darwin's conclusions regarding the transmutation of species."[18] Nevertheless, Agassiz also fell into Eddy's disfavor:

> In one instance a celebrated naturalist, Agassiz, discovers the pathway leading to divine Science, and beards the lion of materialism in its den. At that point, however, even this great observer mistakes nature, forsakes Spirit as the divine origin of creative Truth, and allows matter and material law to usurp the prerogatives of omnipotence. He absolutely drops from his summit, coming down to a belief in the material origin of man, for he virtually affirms that the germ of humanity is in a circumscribed and non-intelligent egg.
>
> *Science and Health, 549:24-550:2*

It is important here to note that, while in Paris, Agassiz had been a student of Georges Cuvier (1769-1832). Cuvier was a well-known naturalist with expertise in the areas of zoology, geology, taxonomy, and paleontology. (Cuvier is often referred to as the *father of paleontology*.) Cuvier recognized irrefutable proof in fossilized evidence for extinct species. However, because Cuvier did not subscribe to the idea of gradual, adaptive change in the production of new species, he concluded that the creation of new species occurred *de novo* after mass extinction events. As a catastrophist and creationist, Cuvier believed that the Creator-God

[18] From *Mary Baker Eddy Mentioned Them,* The Christian Science Publishing Society, Boston, Massachusetts, 1961, p. 10.

repopulated the Earth with some, but not all, previously existing species as well as additional, new species after each mass extinction event that He orchestrated. Thus, Cuvier was neither a proponent of the Genesis account of creation nor a proponent of the then-held view on evolution, which view was specifically Lamarckian and not Darwinian. (Information about Lamarck is provided in Section 2.5.4.4 — entitled *On the Relevance of Lamarck and Haeckel.*)

Eddy used the word *chemicalization* multiple times in her literary works, which word is helpful to the present author in his metaphysical description of the earliest stages in the origin of physical life. To be sure, there are differences between the dictionary definition, Eddy's definition, and the present author's definition for *chemicalization.* A dictionary definition for *chemicalization* is "the act or process of using chemicals." Eddy's definition is "the process which mortal mind and body undergo in the change of belief from a material to a spiritual basis" *(Science and Health, 168:32-169:2).* In contrast to both the dictionary and Eddy, the present author defines *chemicalization* as "the impetus given by the Creator-God to aggregate atoms, ions, compounds, and molecules together in the primordial sea in order to form the organic building blocks necessary for the origin of physical life."

Chemicalization is used by the present author instead of the godless word *abiogenesis,* which term in natural science refers to "the theory that the earliest life forms accidentally developed from inanimate matter." (As a side note here, to describe a word as *godless* is not intended to be pejorative; it just indicates that the Creator-God has not been given a place in its historical meaning.) To be sure, the word *abiogenesis* in traditional evolutionary theory is always used in a *godless* sense. As used here, *chemicalization* includes the first steps in the crystallization of the Creator-God's thinking in the origin of biological life on the planet Earth. (Additional information on *chemicalization* is given in Section 3.1.4.1 — entitled *Chemicalization of Precursors Necessary for Biological Life).*

Although Eddy did not use the term *abiogenesis,* she wrote against what she thought was an illusory process when she stated: "From mortal mind [i.e., the source of all corporeal illusions] comes the reproduction of the species, — first the belief of inanimate, and then of animate matter" *(Science and Health, 189:25-27, brackets mine).* As an interesting side note, Eddy's fault-ridden concept of the female reproductive cell (i.e., an ovum) was greatly influenced by the fanciful, microscopic descriptions of it by Agassiz.

2.5.3.1 Eddy's Cosmology

Eddy's understanding of the universe was predicated on her view that "God's universe is spiritual and immortal" *(Science and Health, 289:24)* and that "the corporeal senses are the only source of evil or error" *(Ibid., 489:24-25).* Yet Eddy's presentation of the universe in her literary works was still enigmatic because, although she tried to ignore matter as nothing (in her words, "no thing"), she really did not succeed. *For example,* Eddy stated that "astronomical order imitates the action of divine Principle; and the universe, the reflection of God, is thus brought nearer the spiritual fact, and is allied to divine Science as displayed in the everlasting government of the universe" *(Ibid., 121:28-32).* If, in Eddy's view, everything corporeal is inclusive of illusion, then why use anything corporeal to prove what exists in a spiritual reality? If one is going to argue that matter and its properties exist only to the physical senses, then one should not use matter and its properties to illustrate spiritual principles of the Creator-God. As the present author sees it, *the Whole Universe* is only an enigma when one fails to take both of its major components into consideration as realities unto themselves: Neither the spiritually-observable universe nor the physically-observable universe is an illusion to those capable of using metaphysically-stereoscopic vision. For the people who use such vision, it is clear that both universes coexist and have parallel yet overlying realities.

Eddy acknowledged the opposing geocentric and heliocentric views of the physical universe and gave credit to Copernicus for sorting the truth out concerning the heliocentric view. She stated: "Copernicus has shown that what appears real, to material sense and feeling, is absolutely unreal. Astronomy, optics, acoustics, and hydraulics are all at war with the testimony of the physical senses" *(No and Yes, 6:23-26)*. Unfortunately, Eddy made an unwarranted cognitive leap by using faulty logic in her line of thinking to conclude: "This fact intimates that the laws of Science are mental, not material; and Christian Science demonstrates this" *(Ibid., 6:26-28)*. Here, individual truths have been poorly cobbled together by Eddy to form a patchwork quilt of unrelated concepts.

Concerning stellar and planetary bodies in the physically-observable universe, Eddy stated:

> Advancing spiritual steps in the teeming universe of Mind lead on to spiritual spheres and exalted beings. To material sense, this divine universe is dim and distant, gray in the sombre hues of twilight; but anon the veil is lifted, and the scene shifts into light. In the record [Eddy was referring to *Genesis 1:23*], time is not yet measured by solar revolutions, and the motions and reflections of deific power cannot be apprehended until divine Science becomes the interpreter. [brackets mine]
>
> *Science and Health, 513:6-13*

The previous quote gives the student of Christian metaphysics the first glint that Eddy may have been seeing the physical universe as a perception-altered form of the spiritual universe and, unlike the present author, not as two separate creations or universes. Eddy implied that what the corporeal senses are witnessing is an elided view of what the present author would call *two parallel realities*. Concerning space, Eddy stated:

> The three great verities of Spirit, omnipotence, omnipresence, omniscience, — Spirit possessing all power, filling all space, constituting all Science, — contradict forever the belief that matter can be actual.
>
> *Science and Health, 109:32-110:3*

> We bury the sense of infinitude, when we admit that, although God is infinite, evil has a place in this infinity, for evil can have no place, where all space is filled with God.
>
> *Science and Health, 469:21-24*

> Divine Science, the Word of God, saith to the darkness upon the face of error, "God is All-in-all," and the light of ever-present Love illumines the universe. Hence the eternal wonder, — that infinite space is peopled with God's ideas, reflecting Him in countless spiritual forms.
>
> *Science and Health, 503:12-17*

Throughout her literary works, Eddy eschewed pantheism. Therefore, it is clear in Eddy's descriptions of *space* that she was describing the *noumenon* of space, or space-in-itself, and not the phenomenon of the empty vacuum of space found in the physically-knowable universe. Again, as was the case for stellar and planetary bodies, Eddy described what she could see of the spiritual universe using her understanding of Christian metaphysics and her corporeal view of the physical universe to work backwards to the spiritual universe — which is to say, Eddy regressed to the Creator-God's original creation and described what she could see in its unfallen state of being. For the present author, this is confirmed in Eddy's statements that: (1) "in divine Science, the universe, including man, is spiritual, harmonious, and eternal" *(Science and Health, 114:27-29);* (2) "the term Science, properly understood, refers only to the laws of God and to His government of the universe, inclusive of man" *(Ibid., 121:28-32);* and (3) "the universe is filled with spiritual ideas, which He evolves, and they are obedient to the Mind that makes them"

(Ibid., 295:6-8). If one can conclude that the Creator-God evolves spiritual ideas in the spiritually- or metaphysically-observable universe, then one's conclusion is not so far afield from the possibility that the Creator-God can also evolve spiritual ideas that manifest as physical objects in the physically-observable universe.

In seeing the spiritually- or metaphysically-observable universe, Eddy had a glimpse of the truth, but that glimpse caused her to put blinders on relative to the reality of the physically-observable universe. Because she had concluded that "the human mind and body are myths" *(Science and Health, 150:32-151:1)* and that "Spirit and its formations are the only realities of being" *(Ibid., 264:20)*, she could not help but conclude that "the physical universe expresses the conscious and unconscious thoughts of mortals" *(Ibid., 484:13-14)*. To Eddy, the existence of matter was only a "supposition of error" *(Ibid., 503:11)* rather than a self-evident, or *a priori,* proposition.

Eddy wanted metaphysics to take the place of physics. The present author desires that metaphysics be used alongside of physics — at least while souls are in corporeality. For Eddy, only the spiritual universe existed. She did not understand, or care to acknowledge, the role of the Luciferian Fall in the alteration of the spiritual universe to produce the physical universe. For this reason, Eddy confused or misunderstood the timeline for "a new heaven and a new earth" *(Revelation 22:1)*. The "new heaven" and "new earth" referred to in Revelation 22:1 only appear: (1) at the end of the millennial reign of peace on Earth by Christ Jesus (i.e., *the Millennium);* and (2) after Christ Jesus has turned the reins of the physical universe over to God the Father *(1 Corinthians 15:28)*. Eddy assumed that what the Apostle John reported as "a new heaven and a new earth" would be seen by all of us immediately upon our making the transition from human life to heavenly life *(Science and Health, 572:19-25)*. To be sure, in her own way, Eddy acceded to the existence of a physical universe, but she purposely refrained from openly admitting it.

According to the present author, upon the introduction of iniquity into the spiritual universe, a modicum of the substance, or essence, of spiritual life was altered to appear as inanimate matter and (eventually) animate matter, but Eddy failed to see this clearly during the majority of her lifetime. The present author does not fault Eddy for this failure because he recognizes the necessity for a gradual progression in Christian metaphysical thinking before one can arrive at the concept of *intelligent evolution*. Although one might be poised to take one's *next step,* one simply cannot take a *next step* before or during one's first step. (Each step that we take is either a *next step* in the right direction or a *misstep* in the wrong direction.) To be sure, the same is true for writing: If we do not write down our first idea on a topic or read it clearly somewhere else, we really are not ready to develop that idea in order to receive the next related idea. *For example,* if the present author had not read and comprehended Eddy's definition of a *year* as a "space for repentance" *(Science and Health, 298:20)*, he would not have apprehended the idea that *time itself constitutes the space for repentance of sins* — or its more succinct version appropriate for the paradigm of *intelligent evolution* — that is, *relative time constitutes relative space*. Thus, in the physically-knowable universe, relative space and relative time are not only merged in fact but also in purpose.

Had Eddy lived longer, it would not have surprised the present author if she had successfully penned an additional book that touched on some of the concepts and ideas covered in *Intelligent Evolution*. Indeed, Eddy was too gifted not to eventually see the multidimensionality of what the present author calls *the Whole Universe* — regardless of the specific ideas or terminology used. The present author certainly acknowledges Eddy's contribution to his own spiritual development in Christian metaphysical thinking. Without her literary works, this work on *intelligent evolution* would never have been attempted or accomplished. In fact, the present author considers his book the *next following step* in Christian metaphysics after Eddy.

Insights granted to the present author have permitted him to understand that, in her later years, Eddy had personal assistants and board members that functioned as enforcers and filters who would have destroyed any unusual written work by her on the topic of Christian metaphysics in order to preserve and protect the organizational bureaucracy of the Church of Christ, Scientist. Had she lived longer, the present author believes that Eddy's final published work would have been entitled *The Altered Universe*.

Eddy believed that "spiritual evolution alone is worthy of the exercise of divine power" *(Science and Health, 135:9-19)*, but the present author would respond that Eddy was not simultaneously attending to both major parts within *the Whole Universe* and, for that reason, was not also considering physical evolution as worthy of the exercise of divine power. Although Eddy did not have physical evolution in mind when she wrote the following, there is really nothing within her statement that is at odds with the concept of *intelligent evolution:* "There is but one primal cause. Therefore there can be no effect from any other cause, and there can be no reality in aught which does not proceed from this great and only cause" *(Ibid, 207:20-23)*. In other words, from the present author's perspective, our Creator-God is not only the primal cause of spiritual evolution through consciousness expansion and unfoldment but also the primal cause of physical evolution through cosmogenesis, abiogenesis, biogenesis, and speciation. Our Creator-God is the only *First Cause, Final Cause,* and *Prime Mover*. Regardless of Lucifer's Fall and the Adamic Fall, our Creator-God is the one true and only real *Cause*. To be sure, cosmological, geological, and biological observations can be interpreted both physically and spiritually at the same time; the two interpretations are not mutually exclusive nor are they meant to be separate and contradictory when one has a metaphysically-stereoscopic view of *the Whole Universe* in Christ Jesus.

2.5.3.2 An Oddity Explained

It is odd to the present author that Eddy wrote that "the perpetuation of the floral species by bud or cell-division is evident..." *(Science and Health, 68:23-25)* because such an acknowledgement would normally lead one to at least speculate that internal changes might occur in a cell (or fertilized ovum) that could account for alterations in an organism's own morphology or in the morphology of the organism's progeny. (Although the words *microevolution* and *macroevolution* would not be coined during Eddy's lifetime, she could have concluded the processes they name without having those two terms available to her.) Regardless of what she could have concluded, there was no knowledge of DNA and the genetic code during Eddy's lifetime — a viable model for DNA not described by Watson and Crick until 1953, forty-three years after Eddy's death.

Because she did not have a knowledge of contemporary biochemistry, genetics, cytology, and mutations, it is comprehensible why Eddy's metaphysics led her to pit matter against divine Mind. (Eddy's *divine Mind* is what the present author refers to as the *Supraconsciousness of the Creator-God.*) She did not know any better because there simply was no better for her to know. A most unfortunate set of circumstances now exists for Christian Scientists of the twenty-first century, who are left with a knowledge of the universe, the planet Earth, and human physiology that dates no later than 1910, the year that Eddy died and the year of her final edition of *Science and Health.* Remember, to Christian Scientists, Eddy is their "forever Leader." For this reason, the majority of practicing Christian Scientists are unable to grow beyond what was known to Eddy during her lifetime. This presents a conundrum as detrimental to spiritual growth as if people only subscribed in thought to what Aristotle or Kant understood and wrote about or to what Pearson understands and writes about.

In *Volume Two,* we will explore the metaphysics of Pierre Teilhard de Chardin as it relates to the concept of *intelligent evolution.*

Afterword to Volume One

In summary at this juncture, metaphysics is the scientific study of invisible reality. The invisible reality studied in metaphysics is purely intellectual, purely spiritual, or a combination of both. For people who have not yet accepted, or who consciously reject, Christ Jesus as the *only-begotten* Son of God, only Savior of the world, and one's only personal Savior, metaphysics can only remain a philosophical endeavor based solely on what human understanding and reason provide and what the intellectual agency of human consciousness enables. For people who accept Christ Jesus as the *only-begotten* Son of God, only Savior of the world, and personal Savior, metaphysics transcends philosophical boundaries to open doors to the unseen in spiritual knowledge from the Holy Spirit's agency of soul-consciousness. For saved human beings, Christian metaphysics possesses both intellectual and spiritual components.

Because saved souls in corporeality are within the earth plane of consciousness, they are exposed simultaneously to different streams of consciousness and different currents within those streams. However, what they are exposed to depends largely on what captures and maintains their personal interest, attention, focus, and commitment. Human beings have access to streams of: (1) spiritually-enblackened consciousness, (2) intellectual consciousness, and (3) spiritually-enlightened consciousness. At any given moment, human beings individually choose to go with the flow from one of the following two stream sets: (a) intellectual consciousness and spiritually-enblackened consciousness; or (b) intellectual consciousness and spiritually-enlightened consciousness.

Without the Creator-God's Holy Spirit residing within one's soul, metaphysics can never be more than a purely intellectual endeavor. In a purely intellectual endeavor, the human brain relies solely on its own cognitive abilities to fill in the blanks concerning unknown information. When a person is educated, literate, and intelligent, the human brain can easily fill in missing information with highly plausible possibilities based on prediction from the person's prior experiences. However, the information supplied may not be accurate. In contrast, when a person has accepted Christ Jesus as the *only-begotten* Son of God, the only Savior of the world, and one's only personal Savior, the Creator-God's Holy Spirit will help to fill in what is missing with accurate information — provided it is the Will of the Creator-God for that person (and other people through that person) to know and understand the missing information at that particular time.

Although information supplied through the agency of the Creator-God's Holy Spirit is always accurate, sometimes the recipient misinterprets the supplied information or provided insight because of the recipient's reasoned conjecture, skewed emotional bias, diminished interest in the topic at hand, or doubt due to a lack in faith. And, although the information that the Creator-God's Holy Spirit supplies may be easy to understand by the recipient because what is being conveyed is simple and direct, it can, instead, be difficult for the recipient to understand because what is being conveyed is complex and multi-layered.

Based on his own experiences, the present author believes that, when the information that the Holy Spirit relays is complex and multi-layered, it is because the Holy Spirit wants the recipient of such information to gradually comprehend specific spiritual concepts over a span of time and not immediately. Why? The Holy Spirit wants the recipient to cogitate on, ponder about, and reflect on certain spiritual truths in order that the spiritual truths conveyed are indelible in — and, therefore, not easily erased from — the recipient's memory. The Holy Spirit wants us not only to grasp the spiritual truths that are being conveyed but also to never let go of grasping them. The Holy Spirit wants the information conveyed to be spiritually savored by the recipient and ever-present for practical application. Understood in these ways, it should be clear to the reader or listener that a perfect understanding of a spiritual truth is generally hard-fought and hard-won but also gratefully received and much appreciated.

Although the Creator-God's Holy Spirit depends on no human being for direction, the Holy Spirit knows that, in supplying missing information or providing insight to a human recipient, at least some of that recipient's experiences and knowledge base must be used to couch the information or insight in order for the recipient to more ably extract meaning from, and practical application for, the information supplied or insight provided. Thus, the Creator-God's Holy Spirit often uses a combination of the recipient's human intelligence and spiritual understanding when providing information or insight to the recipient. The previous statement reiterates that the invisible reality studied in Christian metaphysics is both intellectual as well as spiritually-enlightened for saved human beings. The present author adds that an element of caution is inherent in the explanation just given because of the Holy Spirit's sovereignty: In other words, the Holy Spirit can do and will do whatever the Holy Spirit chooses to do — regardless of what truth the present author or any other author might perceive or how that truth might be presented.

How do you know when information is supplied primarily by your own intellect, and how do you know when information is supplied primarily through the agency of the Creator-God's Holy Spirit?

Let us consider the following example to compare and contrast how the human intellect and the Creator-God's spiritual agency work differently (albeit, at times, complementarily) in supplying missing information and providing insight:

If presented with the sentence "When in _______, do as the ______ do," your intellect may supply the missing words *Rome* and *Romans* if you have heard or read that statement before. If you have never heard or read that statement, your intellect will come up with at least one set of plausible possibilities — *for example,* "When in *danger*, do as the *wise* do." In contrast, the agency of the

Creator-God's Holy Spirit may minister a special insight to you by emphasizing what it is that you: (1) might not already know; or (2) might already know but not necessarily in the same way that the Holy Spirit presents it. *For example,* the Creator-God's Holy Spirit might minister the following statement to the recipient: "When in *Christ,* do as the *righteous* do." (Of course, the Creator-God's Holy Spirit might supply an entirely different structure for a spiritual idea, construct, or concept — which is to say, in a way that departs from utilizing a simple *fill-in-the-blanks* stem or frame.) To be sure, contextual relevancy as well as level and degree of applicability help to determine whether the information supplied or the insight provided is primarily from your own intellect or the agency of the Creator-God's Holy Spirit.

At this juncture, it is important for the reader or listener to learn that, after personally receiving salvation through Christ Jesus and sharing the gospel message of salvation with others, the present author's *raison d'être,* or purpose for earthly existence, is to elucidate Christian metaphysics through the leading of the Creator-God's Holy Spirit for people living during the millennium of peace (i.e., *the Millennium*) — when Christ Jesus reigns for one thousand years on Earth.

The present author believes that, once spiritual truth is said, it can never be unsaid. In other words, as soon as spiritual truth is discovered and articulated in the earth plane of consciousness, it becomes accessible to others who are also in that plane. That is one reason why the same so-called *original* idea can be expressed by two or more different sources in two or more different locations at approximately the same time. And that is why, although human beings can possess spiritual truth, they can never really keep it to themselves or prevent others from accessing it. Thus, a spiritual breakthrough for one person often constitutes a breakthrough for others. Just as one cannot really hide gold or oil in the physical world after it is discovered, so also one cannot really hide spiritual truth in the earth plane of consciousness after it is apprehended. To be sure, the readiness of humanity to receive a revealed truth also plays an important role.

Bibliography: Volume One

Aristotle. *Metaphysics: Translated with an Introduction by Hugh Lawson-Tancred.* New York: Penguin Books, 2004.

Aristotle. *Physics: A New Translation by Robin Waterfield.* Oxford: University Press, 2008.

Berry, George Ricker. *The Interlinear Literal Translation of the Hebrew Old Testament.* (Reprinted from the 1897 Edition). Grand Rapids: Kregel Publications, 1979.

Bullinger, E. W. *Figures of Speech Used in the Bible.* (Reprinted from the 1898 Edition). Grand Rapids: Baker Book House, 1968.

Bullinger, E. W. *The Companion Bible (Facsimile Edition).* Grand Rapids: Kregel Publications, 1922.

Dickey, Adam H. *Memoirs of Mary Baker Eddy.* Santa Clarita: The Bookmark, 2002.

Eddy, Mary Baker. *Concordance to Science and Health with Key to the Scriptures.* Boston: Trustees under the Will of Mary Baker G. Eddy, 1933.

Eddy, Mary Baker. *Complete Concordance to Miscellaneous Writings and Works other than Science and Health.* Boston: Trustees under the Will of Mary Baker G. Eddy, 1915.

Eddy, Mary Baker. *Prose Works other than Science and Health with Key to the Scriptures.* Boston: The First Church of Christ, Scientist, 1953.

Eddy, Mary Baker. *Science and Health with Key to the Scriptures.* Boston: Christian Science Board of Directors, 1906.

Haushalter, Walter M. *Mrs. Eddy Purloins from Hegel.* Boston: A. A. Beauchamp, 1936.

Hawking, Stephen. *A Brief History of Time.* New York: Bantom Books, 2017.

Hawking, Stephen. *Brief Answers to the Big Questions.* New York: Bantom Books, 2018.

Hegel, Georg Wilhelm Friedrich. *Die Phänomenologie des Geistes (1807).* Project Gutenberg eBook, 19 June 2012 <www.gutenberg.org/catalog/world/readfile?fk_files=1464225>.

Kant, Immanuel. *Prolegomena to any Future Metaphysics Which Will Be Able to Come Forth as Science (translation from the 1783 edition).* New York: The Liberal Arts Press, 1950.

Kant, Immanuel. *Prolegomena zu einer jeden künftigen Metaphysik die als Wissenschaft wird auftreten können (Erstdruck: Riga 1783).* Berliner Ausgabe, 2 Auflage, 2013.

Kant, Immanuel. *The Metaphysical Foundations of Natural Science.* Lexington: Translated by Ernest Belfort Bax, 2015.

Levinson, Gene. *Rethinking Evolution: The Revolution That's Hiding in Plain Sight.* World Scientific: New Jersey, 2019.

Mary Baker Eddy Mentioned Them. Boston: The Christian Science Publishing Society, 1961.

Pearson, Joseph Adam. *As I See It: The Nature of Reality by God.* Dayton: Christ Evangelical Bible Institute, 2015.

Quimby, Phineas Parkhurst. *The Quimby Manuscripts (1846-1865).* London: Forgotten Books (Classic Reprint Series), 2015.

Strong, James. *Strong's Exhaustive Concordance of the Bible.* Nashville: Crusade Bible Publishers, Inc., 1890.

Walvoord, John F. and Roy B. Zuck (editors). *The Bible Knowledge Commentary: An Exposition of the Scriptures by Dallas Seminary Faculty [New Testament Edition]*. Elgin: David C. Cook, 1983.

Walvoord, John F. and Roy B. Zuck (editors). *The Bible Knowledge Commentary: An Exposition of the Scriptures by Dallas Seminary Faculty [Old Testament Edition]*. Colorado Springs: Cook Communications Ministries, 2000.

Vine, William E., Merrill F. Unger, and William White. *Vine's Complete Expository Dictionary of Old and New Testament Words*. Nashville: Thomas Nelson, Inc., 1985.

Webster, Noah. *Noah Webster's First Edition of An American Dictionary of the English Language (Facsimile Edition)*. Anaheim: Foundation for American Christian Education, 1967.

Zondervan Parallel New Testament in Greek and English. Grand Rapids: Zondervan Bible Publishers, 1975.

Books by the Author

As I See It: The Nature of Reality by God by Rev. Joseph Adam Pearson, Ph.D., Christ Evangelical Bible Institute, Copyright 2022. ISBN 978-0615590615.

Classroom Version of As I See It: The Nature of Reality by God by Rev. Joseph Adam Pearson, Ph.D., Christ Evangelical Bible Institute, Copyright 2022. ISBN-13: 978-1734294705.

God, Our Universal Self: A Primer for Future Christian Metaphysics by Rev. Joseph Adam Pearson, Ph.D., Christ Evangelical Bible Institute, Copyright 2020. ISBN 978-0985772857.

Divine Metaphysics of Human Anatomy by Rev. Joseph Adam Pearson, Ph.D., Christ Evangelical Bible Institute, Copyright 2021. ISBN 978-0985772819.

Hello from 3050 AD! by Rev. Joseph Adam Pearson, Ph.D., Christ Evangelical Bible Institute, Copyright 2021. ISBN 978-0996222402.

Christianity and Homosexuality Reconciled: New Thinking for a New Millennium! by Rev. Joseph Adam Pearson, Ph.D., Christ Evangelical Bible Institute, Copyright 2021. ISBN 978-0985772888.

The Koran (al-Qur'an): Testimony of Antichrist by Rev. Joseph Adam Pearson, Ph.D., Christ Evangelical Bible Institute, Copyright 2020. ISBN 978-0985772833.

Telugu Version of Quran: Testimony of Antichrist by Rev. Joseph Adam Pearson, Ph.D., Christ Evangelical Bible Institute, Copyright 2020. ISBN 978-0996222457.

Urdu Version of Quran: Testimony of Antichrist by Rev. Joseph Adam Pearson, Ph.D., Christ Evangelical Bible Institute, Copyright 2021. ISBN 978-0996222440.

Revelation of Antichrist by Rev. Joseph Adam Pearson, Ph.D., Christ Evangelical Bible Institute, Copyright 2021. ISBN 9780996222488.

Intelligent Evolution by Rev. Joseph Adam Pearson, Ph.D., Christ Evangelical Bible Institute, Copyright 2022. ISBN 978-0996222426.

The Biology of Psychism from a Christian Perspective by Rev. Joseph Adam Pearson, Ph.D., Christ Evangelical Bible Institute, Copyright 2020. ISBN 978-0996222464.

The Threeness of God by Rev. Joseph Adam Pearson, Ph.D., Christ Evangelical Bible Institute, Copyright 2021. ISBN 978-1734294729.

Feel free to contact the author at:

DrJPearson@aol.com
or
drjosephadampearson@gmail.com

CHRIST EVANGELICAL BIBLE INSTITUTE

INTELLIGENT EVOLUTION

VOLUME TWO

Rev. Joseph Adam Pearson, Ph.D.

Copyright © 2022 by Rev. Joseph Adam Pearson, Ph.D.

All rights reserved.

This work is a revision of the following earlier versions
registered with the United States Copyright Office:

Copyright © 2021 (TX0009070948) by Joseph Adam Pearson
Copyright © 2020 (TX0008869477) by Joseph Adam Pearson
Copyright © 2018 (TX0008566448) by Joseph Adam Pearson
Copyright © 2017 (TX0008361069) by Joseph Adam Pearson
Copyright © 2016 (TX0008239803) by Joseph Adam Pearson

Hardcover Identifier:
ISBN-13: 9781734294774

Published by
Christ Evangelical Bible Institute
(SAN: 920-3753)
Dayton, Tennessee

Last edited on May 21, 2022

VOLUME TWO
Table of Contents

An Introduction to Volume Two

The Importance of Christ Jesus to Creation, Re-Creation, Evolution, Expansion, and Consciousness Unfoldment

Christ Jesus, the Only-Begotten of Yahweh

μονογενής (pronounced mo-no-ge-nase´) [G3439] is the most important word in the Greek New Testament when it is used in conjunction with the physical conception and birth of Christ Jesus as the Son of Yahweh, the God of the Holy Bible. **μονογενής** (mo-no-ge-nase´) is a compound word composed of the two base words **μόνος** (pronounced mo-nos´) [G3441] and **γεννάω**[19] (pronounced gen-au´) [G1080]. **μόνος** (mo-nos´) means: *one, only, only one, one and only, solitary, and unique;* and **γεννάω** (gen-au´) means: *born* (i.e., delivered from a uterus), *begat, begotten, birthed, conceived, generated, and legitimate.* Because **μονογενής** (mo-no-ge-nase´) is a compound word, its complete meaning includes the individual meanings of both root words and not just the meaning of one of them. In other words, the full definition for **μονογενής** (mo-no-ge-nase´) includes: *only-begotten, one and only physically born, only legitimate, uniquely-conceived,* and *solitarily-generated.* Although some scholars have chosen to define **μονογενής** (mo-no-ge-nase´) by the single word *only* because they believe that the *begotten* portion is redundant, implied, archaic, and/or unrelatable to the

---

[19] γεννάω (pronounced gen-au´) [G1080] is a variation of γένος (pronounced gen´-os) [G1085], which is derived from γίνομαι (pronounced gē´-no-mī) [G1095]. γένος (gen´-os) means: *genus, kind, kindred, offspring, descendant* (i.e., child), or *species*; and γίνομαι (gē´-no-mī) means *originated* or *ordained*.

modern ear, the definition *only* without *begotten* is, in fact, an under-translation because it is missing one-half of its full meaning. To be sure, using the single word *only* to define μονογενής (mo-no-ge-nase′) does not impart the same meaning as using *only-begotten.*

Many people do not grasp the meaning of *begotten* in the expression *only-begotten Son of God.* Therefore, for the sake of clarity, it is important to state here that "begotten" is derived from Old English and is the past participle of the verb "beget," whose past tense is "begat" *(beget, begat, begotten).* The word *beget* means "to bear" *(bear, bore, born),* "to give birth to," and "to produce offspring." Thus, the word "begotten" means "born," "birthed," "conceived," or "physically delivered from a uterus." The first man Adam was not "begotten" by the God of the Holy Bible because the first man Adam was neither from a fertilized egg nor delivered from a uterus and because the first man Adam was neither self-existent nor equivalent to the Creator-God. Christ Jesus, however, is self-existent and equivalent to the Creator-God *(John 1:1).* Only Christ Jesus was "the begotten" of God. Although the first man Adam was "the Son of God" *(Luke 3:38 KJV),* the first man Adam was a created being and never God-in-flesh (i.e., *God Incarnate)* as was Christ Jesus *(John 1:14 KJV).*

In the case of Christ Jesus, "begat by God" and "begotten by God" mean: (1) that God Himself provided the seed and Mary (Miriam) herself provided the egg for the conception of Christ Jesus; (2) that Christ Jesus was physically delivered from Mary's uterus; and (3) that Christ Jesus was composed of the same spiritual substance as God in addition to human flesh. Christ Jesus was not generated through sexual relations but through the Creator-God's Holy Spirit *overshadowing* Mary *(Luke 1:35 KJV).* Mary the mother was a full participant in the conception and birth of Christ Jesus through her personal physical contributions of egg, uterus, and placental nutrition. (Mary was not just an incubator into which a *second Adam* had been placed.) Although Yahweh is the Father and Mary is the mother of Christ Jesus, and Christ Jesus is God-in-flesh,

Mary is neither *the wife of God* nor *the mother of God.* Christ Jesus was the unique hybrid of the Creator-God's Holy Spirit and Mary's corporeality (i.e., her physical substance).

Although the first man Adam was *created* in the complete image and perfect likeness of the Creator-God, the first man Adam was not equal to the Creator-God. In other words, the first man Adam was not God. In contrast, Christ Jesus was composed of the same self-existent substance as God and, as such, is uniquely one in being, or spiritually conjoined, with the Creator-God. Thus, Christ Jesus was, is, and always will be the same as God because he, in fact, is God Himself. The first man Adam was made only of *created* substance; in contrast, Christ Jesus was composed of the same self-existent substance as God that uniquely appeared in physical flesh. (Christ Jesus was, is, and always will be God regardless of the state or condition of being that he was, is, or will be in.)

In order to define *only-begotten* correctly concerning Christ Jesus, it is important to properly contextualize μονογενής (mo-no-ge-nase´) according to the writings of John the Apostle. Christ Jesus is known as "the Word of God" not only in Revelation 19:13, written by John the Apostle, but also in the Gospel According to John:

> {1} In the beginning was the Word, and the Word was with God, and the Word was God. {2} The same was in the beginning with God. {3} All things were made by him; and without him was not anything made that was made. {4} In him was life; and the life was the light of all people. {5} And the life's light had shone in darkness, but darkness could not comprehend it. {10} He was in the world, and the world was made by him, but the world did not recognize who he was. {14} And the Word was made flesh, and dwelt among us, and we beheld his glory — the glory as of the *only begotten* [μονογενής] of the Father, full of grace and truth. {18} No one has seen God at any time; the *only begotten* [μονογενής] Son, who is at the core of the Father, he has declared Him. {34} And I [John] saw him, and bare record that he is the Son

of God. {49} And Nathanael responded to Jesus and said, "Rabbi, you are the Son of God; you are the King of Israel." [italics and brackets mine]

John 1:1-5, 10, 14, 18, 34, & 49 KJV Paraphrase

To summarize at this juncture:

1. Christ Jesus is "the Word of God" *(Revelation 19:13 KJV)*.
2. The Word of God is the Creator-God *(John 1:1 KJV)*.
3. The Word was made flesh as the *only-begotten* Son of God in Christ Jesus *(John 1:14, 18, 34, and 35 KJV)*.
4. Christ Jesus is *God Incarnate (i.e., God-in-flesh)*.
5. God was in the world that he had made, but the world did not recognize him because those in darkness could not see his transfigured glory — the glory that John, James, and Peter had witnessed on the mountain when Jesus spoke with Elijah and Moses *(John 1:4, 5, 10, and 14 KJV)*.

For as long as people on Earth consciously reject Christ Jesus as the *only-begotten* Son of Yahweh and God-in-flesh, they place themselves under the curse of Yahweh's Wrath, or His Justified Anger, not only while they are on Earth but also throughout eternity. However, for the duration that souls are in corporeality (i.e., in a human body), they still have an opportunity (not necessarily just *one* opportunity) to remove themselves from the curse of Yahweh by: (1) accepting Christ Jesus as the *only-begotten* Son of Yahweh and God-in-flesh *(God Incarnate);* and (2) accepting his sacrifice on the cross of Calvary as the only sacrifice acceptable to God the Father for the atonement of their iniquity and sins and the remission of the debt they owed for their iniquity and sins.

Satan and his demons do not mind if people accept that: (1) Christ Jesus is one prophet of many prophets; (2) Christ Jesus was born of the virgin Mary (Miriam); (3) Christ Jesus is the prophesied Messiah of Israel; (4) Christ Jesus was a worker of miracles; and (5) Christ Jesus will return one day to defeat the Antichrist, or False Messiah. However, Satan and his demons are adamant that no one on Earth learn that: (1) Christ Jesus is the *only-begotten* Son of God; (2) Christ Jesus is the only incarnation of God in human flesh; (3) Christ Jesus is the Savior of the world through the shedding of his blood as he was dying on the cross at Calvary; and (4) Christ Jesus is our only personal Savior — all four concepts explicitly and implicitly stated and restated in the New Testament.

By influencing human beings to reject the four concepts given in the previous paragraph, Satan and his demons help confirm for Christians that, of all theological concepts, these four concepts are the most powerful for people on Earth to know. How do we know that they are the most powerful? They threaten Satan in his mission to prevent the salvation of human beings and, thereby, seek to rob Yahweh of the restoration of His creation. To be sure, although Satan is the enemy of all human beings, Satan is only our indirect enemy; Satan's true Enemy is Yahweh, the God of the Holy Bible. It is for this reason that Satan seeks to rob Yahweh of His creation. Satan erroneously believes that, by robbing Yahweh of His creation, Satan will unseat Yahweh as Supreme Being and replace Him as universal Sovereign.

Everything that Satan has done after his fall has been to fulfill his desire of robbing the Creator-God of His creation, unseat the Creator-God as Supreme Being, and replace the Creator-God as universal Sovereign. To this end, throughout the history of humankind, Satan has tried to: (1) murder all Jews, (2) murder all Christians, (3) discredit the witness of Jews and Christians, (4) firmly establish the Antichrist religion of Islam throughout the Earth, and (5) cause all people on Earth to doubt the accuracy of the Old and New Testaments and the validity of the gospel message of salvation through Christ Jesus as the *only-begotten* Son of God and God-in-flesh.

The central doctrines of Islam promote that: (1) Christ Jesus is not the *only-begotten* Son of God (Islam teaches that it is blasphemy to say that Allah has an *only-begotten* Son); (2) Christ Jesus was not God-in-flesh (Islam teaches that it is a blasphemy to say that Christ Jesus is God the Son); and (3) Christ Jesus is not an intercessor between human beings and God (Islam teaches that Christ Jesus did not really die on the cross and that the Islamic Allah does not need an intercessor for the salvation of human beings if the Islamic Allah desires to save them).

To minimize the significance of the word *begotten* when used about Christ Jesus, literate Muslims sometimes refer to Psalm 2:7 from the King James Version of the Bible to say that, according to the Bible, King David was also "begotten" of God. It states in Psalm 2:7 (KJV): "I will declare the decree: the Lord God Almighty has said to me, 'You are my Son; this day have I *begotten* you [italics mine].'" However, in Psalm 2:7, the God of the Holy Bible is not speaking about King David but about the *King of Kings,* who is Christ Jesus. In other words, the "begotten" in Psalm 2:7 is Christ Jesus. To be sure, the entire Second Psalm is prophetic Scripture about Christ Jesus and not King David. (There are many verses throughout the Book of Psalms that are prophetic concerning Christ Jesus.) It states clearly in Psalm 2:8 (KJV): "'Ask of Me [the Lord God Almighty], and I will give you the heathen for your inheritance, and the uttermost parts of the earth for your possession [brackets mine].'" The Hebrew word for "the heathen" (or "the nations" in other translations) is *goyim,* which means "the Gentiles." King David did not rule over the Gentiles throughout the whole world (i.e., to "the uttermost parts of the earth"): King David was the king of the children of Israel in the land of Israel. Only the Savior of the world, Christ Jesus, rules over Gentiles throughout the whole world. To further confirm this understanding, whenever the word *begotten* from Psalm 2:7 is referenced in the New Testament

(Acts 13:33, Hebrews 1:5, and Hebrews 5:5), it is always concerning Christ Jesus and not King David.

In transitioning to the next section, it is important for the reader to always remember that Christ Jesus is the Christian God. *That* is why he is described as the deific Force, divine Principle, creative Logos, and spoken Word responsible for all creation, re-creation, evolution, expansion, and consciousness unfoldment. *That* is why Christ Jesus is worshiped and exalted.

The Transformative Nature of Christ Jesus

Christ Jesus is the deific Force, creative Logos, divine Principle, and spoken Word at the core of the Creator-God — or, using the jargon of the King James Version, "in the bosom of the Father" *(John 1:18 KJV)*. As such, in addition to the characteristics and qualities explained in the immediately preceding section, Christ Jesus is responsible for: (1) all creation; (2) holding together not only each atom but also *the Whole Universe;* (3) all re-creation and restoration to God the Father of all that is His; (4) all cosmic, biological, and consciousness evolution; (5) all phylogeny and ontology (explained in Section 2.5.4.4 — entitled *The Relevance of Lamarck and Haeckel);* and (6) all four stages in our personal spiritual development.

Regardless of the specific verbiage used, Christians living before *the Millennium* should already be aware of the following four developmental stages in the consciousness evolution of individual saved souls in corporeality:

1. Preparation of the individual's emotions, intellect, and will to receive salvation. (Depending on individual, this may be a short term process or a lifelong process.)

2. The actual receiving of salvation itself through: (a) one's personal belief in Christ Jesus as the *only-begotten* Son of God; and (b) one's confession of faith in Christ Jesus as personal Savior. (Belief and confession may occur at the same time or at different times.)

3. Sanctification through perpetual contrition, self-discipline, spiritual focus, and an unwavering desire to please the Creator-God. (This process continues throughout the remainder of one's life on Earth after one receives salvation.)

4. The continuity of one's life in an immortal state of being throughout eternity after the death of one's physical body (provided, of course, that one received salvation while in one's physical body).

Most Christians are not aware that, in addition to the four personal development stages just listed, Christ Jesus' transformative nature is also responsible for: (1) all cosmogenesis, (2) all chemogenesis, (3) all abiogenesis and the emergence of all physical life, (4) all phylogenesis, (5) all embryogenesis, and (6) all consciousness evolution by controlling the directions and movements of all electromagnetic radiation, all gravitational interaction, all quantum mechanics, and every interactive force in the physically-knowable universe. In short, Christ Jesus is the Driver, Prime Mover, Engineer, and Orchestrator behind all activities in the physically-knowable universe with the exception of activities controlled by Evil — Evil consisting of Satan, his fallen angels, and his unclean spirits, demons, evil spirits, or devils.

The Unfortunate Separation of *Christ* and *Jesus*

The First Council of Nicea was convened in 325 AD to establish a common doctrine, creed, and canon law acceptable to the majority of Christendom. The most important work of the First Council of Nicea was in resolving the relationship of *God the Son* to *God the Father* and articulating that relationship in the written Nicene Creed. After the Council of Constantinople in 381 AD, the resulting creed included the refined statement of Christian belief "in one Lord Jesus Christ, the only Son of God, begotten from the Father before all ages, God from God, Light from Light, true God from true God, begotten, not made; of the same essence as the Father."

Students who have a thorough understanding of the Holy Bible recognize that one cannot have "Jesus" without having "Christ" (or "the Christ") and one cannot have "Christ" (or "the Christ") without having "Jesus." Christians should try to hold the whole name, "Christ Jesus" (or "Jesus the Christ"), while they simultaneously attend to its two parts, "Jesus" and "(the) Christ."

The words *Jesus* and *Christ* are inextricably linked together and should rarely be used separately so as not to confuse the hearer, the reader, or even oneself (yes, we can easily confuse ourselves). The English word *Christ* is a title derived from the Greek word **Χριστός** (*Christos*) [G5547] and its counterpart in Latin, *Christus* [CHRISTVS]. The Greek word *Christos* [G5547] is a translation of the Hebrew word *H'Moshiach* [H4899], which means "*the* Messiah" or "*the* Anointed One" in English. And the Greek word **Μεσσίας** *(Messias)* [G3323] is the transliterated form of the Hebrew word *Moshiach* [H4899], meaning "Messiah" or "Anointed One."

Unfortunately, many students of the Holy Bible are not aware that the noun *Christ* has been trivialized by those who entertain certain inaccurate concepts from Eastern religions, New Age philosophy, Theosophy, and Christian metaphysics.

For example, the Hindu deity Krishna, the supposed earthly incarnation of the Hindu deity Vishnu, is worshiped by seeking to propagate his consciousness as revealed in various Hindu scriptures. (For the sake of clarification, the Holy Bible is the one true and only real Scripture.) This concept of *Krishna Consciousness* has been imported to Christianity as an ill-defined *Christ Consciousness.*

To be sure, when used alone, the word *Christ* may accurately imply a spiritual state of mind and a heightened level of consciousness; unfortunately, however, when used alone, the word *Christ* can also inaccurately imply that the spiritual state of mind and heightened level of consciousness can be achieved without accepting the Biblical Jesus as: (1) the *only-begotten* Son of God, (2) the only Messiah of Israel, (3) the one true Savior of the world, (4) one's only personal Savior, (5) the Word of God, and (6) the Christian God. Indeed, one cannot have "the Christ," "the mind of Christ," "divine Mind," "universal Mind," "Christ Consciousness," or "the Supraconsciousness of God" without accepting the shed blood of the *only-begotten* Son of God as the only sacrifice acceptable to God the Father for the remission of our sins and the cancellation of the debt we owe to Him for those sins.

One of the earliest representations of codification in the separation of *Jesus* from *Christ* and *Christ* from *Jesus* is seen in the writings of Phineas Parkhurst Quimby (first referred to in Section 2.5.3 of Volume One in *Intelligent Evolution),* whose errant ideas helped serve to form related false doctrines propagated by such religious movements as *Christian Science, Unity,* and *Religious Science* (i.e., *Science of Mind).* Quimby's separation of *Christ* from *Jesus* is captured in the following quotes from three of his essays:

> Jesus called this [divine and scientific] truth *the Son of God.* Peter called it *Christ.* The people's ignorance confounded the two together and called it *Jesus Christ.* This last construction has given rise to all the religious wars and bloodshed since the Christian era [began]. [brackets and italics mine] *Quimby, Christ and Truth, January 1860, in The Quimby Manuscripts, p. 197*

> I will try to explain the true Christ from the false Christ, and show that "Christ" never was intended to be applied to Jesus as a man, but to a Truth superior to the natural man — and this Truth is what the prophets foretold. *Quimby, True and False Christs, January 1860, in The Quimby Manuscripts, p. 201*

> The idea that the man Jesus was anything but a man, was never thought of. Jesus never had the least idea of such an explanation *[that he should be called the Christ].* [brackets mine] *Quimby, Jesus and Christ, March 1860, in The Quimby Manuscripts, p. 216*

Let it be understood that making the words and meanings of *Christ* and *Jesus* mutually exclusive leads one to gross doctrinal error and, as a result, an overall weakening of what should be one's personal and empowering faith through Jesus Christ.

For the purposes and intents of *Intelligent Evolution,* readers should assume that, when the present author uses the expression *Christ Consciousness,* it is synonymous with *the Consciousness of the Biblical Jesus,* who is *God Incarnate* (i.e., God-in-flesh).

2.5.4 Insights, Implications, and Applications from de Chardin

2.5.4.1 An Introduction to de Chardin's *The Phenomenon of Man*

One of the reasons that we read the written works of other authors is: (1) to see what insights they may have had or may have missed; (2) to see if we can determine the validity or invalidity of those insights (even the ones we think they missed); and (3) to see if we can expand upon those insights with insights of our own. We do not read other authors in order to come to complete agreement with them, think exactly as they do, or express ourselves exactly as they have. We read other authors that we might better learn to think for ourselves. Of such is the case for reading the works of Aristotle, Kant, Eddy, de Chardin, and Hawking with the purpose of thinking about ideas, concepts, and constructs potentially relatable to the paradigm of *intelligent evolution*.

As we seek to understand an author's specific writing or publication, it is important not only to *read the lines* (that is, pay attention to specific word and phrase meanings through grammar, syntax, and parsing) but also to *read behind the lines* by understanding historical contexts and word etymologies (i.e., not only historical word origins but also a particular author's intended — and even peculiar — meanings for certain words) as well as the author's background, biography, and stated intent for his or her writing. And, as we seek to understand an author's specific writing, it is important not only to *read behind the lines* but also to *read between the lines* by learning to identify implications of, and inferences from, the author's writing. And, as we seek to understand an author's writing, it is important not only to *read between the lines* but also to *read beyond the lines* concerning applications of its written truths to related events, circumstances, and realities as well as applications of its ideas, constructs, and concepts to contexts not originally intended by a particular author:

This is what the present author has sought to do with the cited works of Aristotle, Kant, Eddy, de Chardin, and Hawking and their ideas specifically relatable to his paradigm of *intelligent evolution.*

When reading a book by any author, it is important to read all introductory notes by the author and notes by the book's translator (if a translated edition exists). We must also read the book's introduction, preface, foreword, epilogue, postscript, appendix, footnotes, and endnotes. Examples of why this is necessary are found in the remaining paragraphs of this current section (2.5.4.1).

The primary source for the present author's section on de Chardin is de Chardin's work entitled *The Phenomenon of Man (Le phénomène humain).* The translation used by the present author is an edition in English that is very popular, if not the most popular. It is the present author's opinion that the translator of *Le phénomène humain* committed a grave error by choosing to eliminate de Chardin's "initial capitals for all abstract nouns such as 'science,' 'life,' 'thought,' and also for 'world,' 'universe,' 'man' and other such key words of his work" in order to make the printed page look "more normal to the English reader" *(Phenomenon, Translator's Note, page 9).* This is a grave error because it is standard for authors of metaphysical works, even in English, to capitalize nouns that are intended to have abstract and/or transcendent meanings to signal to their readers that the word meanings go well beyond what is normally intended and perceived. Fortunately, the present author had a copy of *Le phénomène humain* in French to ensure that capitalized nouns with abstract or transcendent meanings could be determined and noted for the purpose of writing this entire section (2.5.4).

If readers were unfamiliar with the background and biography of de Chardin and only read the main body of *The Phenomenon of Man,* and not its Epilogue, Postscript, Appendix, and Footnotes, they might misconclude that: (1) de Chardin did not have a personal relationship with Jesus Christ; (2) de Chardin asked philosophic questions without incorporating Jesus Christ into their ultimate answers; and (3) de Chardin put his faith and hope in humankind as the source of all love. However, it is primarily in the Epilogue, Postscript, Appendix, and Footnotes of *The Phenomenon of Man* that de Chardin makes clear his faith in "the uncompromising affirmation of a personal God" through His "Redeeming Incarnation [as Christ Jesus]" *(Epilogue, p. 293, brackets mine)* when he states that "Christ invests himself organically with the very majesty of his creation" *(Ibid., p. 297)*. To be sure, de Chardin comprehended about Christ what many seasoned Christians do not comprehend — which is to say, that Christ Jesus *is* the Creator-God *(see John 1:1-5)*.

In one of the footnotes to his Postscript, de Chardin states:

> For a Christian believer it is interesting to note that the final success of hominisation[20] (and thus cosmic involution) is positively guaranteed by the 'redeeming virtue' of the God incarnate in his creation. But this takes us beyond the plan of phenomenology.
>
> *Postscript, Footnote 2, page 308*

In the Appendix to *Phenomenon*, de Chardin declares in his final statement:

---

[20] ***Hominisation*** **(after the French and British spelling) is a word coined by de Chardin, who explained that it "can be accepted in the first place as the individual and instantaneous leap from instinct to thought, but it is also, in a wider sense, the progressive phyletic spiritualisation in human civilisation of all the forces contained in the animal world"** ***(Phenomenon, p. 180)*****. In other words, concerning the word's broader sense, there is an additive aspect to** ***hominisation*** **that runs the full gamut of consciousness from the level of animalcules up through the entire spectrum of anthropogenesis. Readers of de Chardin's works should not confuse the word** ***hominisation*** **(or** ***hominization*****) with the word** ***humanization*****; these two words are not synonymous. [A further explanation for** ***hominisation/hominization*** **is given by the present author later in this section (2.5.4).]**

> In one manner or the other it still remains true that, even in the view of the mere biologist, the human epic resembles nothing so much as a way of the Cross.
>
> *Appendix, page 313*

The present author cautions readers to be circumspect and judicious concerning reviews or explanations written about an author by someone who has made only a cursory examination of the author's life or of topics about which that author has written. *For example,* de Chardin uses the word *orthogenesis*[21] in *The Phenomenon of Man* to describe a particular theory that he espoused. If the meaning and explanation of the theory named by this word is reviewed in *Wikipedia,* one would find that *orthogenesis* represents hypotheses that are *obsolete, rejected, refuted, collapsed,* and *dead.* Although outdated hypotheses in natural science may be described using those words because the hypotheses can be proven or disproven based on facts determined *a posteriori, a priori* hypotheses in metaphysics can neither be proven nor disproven because they describe an invisible reality that transcends physical reality and, in effect, are untestable except by an uncommon common sense vis-à-vis *pure understanding* and *pure reason.* To be sure, de Chardin defined *orthogenesis* differently than many other scholars who have used or currently use the word. But one would only know that from reading it in contexts used and explained by de Chardin.

It is important to note that de Chardin coined new words and slanged old ones to convey his unique message. It is also important to note that the French and British spelling of many of de Chardin's words are retained by the present author in quotes from the translated edition of *The Phenomenon of Man.*

---

[21] **As used by de Chardin, *orthogenesis* describes the increasing complexification of living matter in evolution: In other words, rather than just spread, living matter ascends in complexity in improbable ways through evolution. (See *Phenomenon of Man, p. 109.*)**

2.5.4.2 The Psychism of de Chardin

Because of the word's association with satanic witchery by most Christians and with commercial charlatanism by most intelligent people, the word *psychic,* as either noun or adjective, is often not a good word to use in 21st century American society. However, people who have a distaste for the word *psychic* are either ignorant of the word's etymology or have chosen to ignore the continued importance of the word and its related variants and derivatives in academic theology, philosophy, and modern science.

Just as the Greek word *pneuma* was important to Aristotle and continues to be important to New Testament scholars, so does the Greek word *psuche* — from which the words *psyche, psychic, psychical, psychism, psychology,* and *psychiatry* derive — continue to have importance in academic theology, philosophy, and modern science.

Based on the two Greek nouns most often used in the New Testament for a human being's "spirit" and "soul" — πνεῦμα *pnyü'-mä* [G4151] ("spirit" or "breath") and ψυχή *psü-khā'* [G5590] ("soul" or "mind") — one might define: (1) *spirit* as "the invisible essence of a human being characterized by his or her unique personality;" and (2) *soul,* or *mind,* as "the seat of a human being's thoughts and feelings that impart his or her consciousness." In modern scientific contexts (especially in biology, neurology, psychiatry, and psychology), ψυχή *psü-khā'* ("soul" or "mind") refers to consciousness that is associated with general regions and specific areas of the human brain. Thus, what is *psychic,* or *psychical,* is in contrast to what is *physical.*

Highly educated biologists, philosophers, theologians, historians, psychologists, psychiatrists, neurologists, and linguists might spend a significant amount of time trying to elucidate and clarify the differences between the "spirit" and the "soul" of a human being. So, we will leave that work to them. However, it is important to add at this juncture that the combined "spirit" and "soul" of a human being are like dissolved sugar in distilled water: The water molecules and sugar molecules are indivisible from one another under normal circumstances and conditions. In other words, the combined "spirit" and "soul" of a human being are indivisible from one another under normal circumstances and conditions. To be sure, one's combined "spirit" and "soul" are, however, distinguishable as well as divisible from one's somatic, or physical, identity even though they both can influence one's bodily form, likeness, appearance, and physiology.

Indeed, ψυχή *psü-khā'* is the Greek word from which the English nouns *psychic* and *psychology* have been derived. Thus, just as the word *psychology* means "the study of the *mind,"* so, in certain contexts, can the noun *psychic* mean "*mind* reader." In an extended sense, the adjective *psychic* goes well beyond describing functions of the cerebrum as detected by human brain wave activity. And, in its most transcendent sense, the adjective *psychic* describes detecting activity in the invisible, electromagnetic, and supernatural realm by gifted and talented "sensitives," "susceptible channels," or "psychics." Metaphysically speaking, to be a *psychic* means that one "receives impressions from the soul, or mind, of another person" and that one "is able to transcend space-time in order to sense consciousness aspects of the past, present, and future." Common 21st century cultural connotations of *psychic* activity include clairvoyance (knowing beforehand), clairaudience (perceiving what is inaudible), far memory ("remembering" events and circumstances outside of one's own lifetime), and telepathy (reading another person's mind).

In de Chardin's *The Phenomenon of Man,* the words *psychic, psychical,* and *psychism* are never used in reference to the common 21st century cultural connotations listed in the last sentence of the preceding paragraph. Rather, *psychic, psychical,* and *psychism* had very broad meanings to de Chardin that included everything and anything associated with consciousness, the *interiority,* or *the Within:*

We shall assume that, essentially, all energy is *psychic* in nature… [italics mine]

Phenomenon, p. 64

It is generally accepted that we must assume psychic life to 'begin' in the world with the first appearance of organized life, in other words, of the cell.

Phenomenon, p. 88

From the moment we regard evolution as primarily psychical transformation, we see there is not one instinct in nature, but a multitude of forms of instincts each corresponding to a particular solution of the problem of life. The 'psychical' make-up of an insect is not and cannot be that of a vertebrate; nor can the instinct of a squirrel be that of a cat or an elephant: this in virtue of the position of each on the [phylogenetic] tree of life. [brackets mine]

Phenomenon, p. 167

Here, and throughout this book *[The Phenomenon of Man]*, the term 'consciousness' is taken in its widest sense to indicate every kind of *psychism,* from the most rudimentary forms of interior perception imaginable to the human phenomenon of reflective thought. [brackets and italics mine]

Phenomenon, Footnote 1, p. 57

Thus, for de Chardin, *psychism* includes cytoplasmic streaming, taxes (i.e., behavioral responses to external stimuli [pronounced tak-seez´]), tropisms, instincts, self-reflections, intuitions, socializations, and consciousness convergence by any and all living things.

From the biosphere to the species, [evolution] is nothing but an immense ramification [branching] of *psychism* seeking for itself [to be expressed] through different forms. [brackets and italics mine]

Phenomenon, p. 151

To de Chardin, *psychism* existed in the protoplasm of the very first primordial cells just as it exists in the physical centers of consciousness (e.g., cerebral hemispheres) of the most complex vertebrates living today. Extremely important to the present author's paradigm of *intelligent evolution,* de Chardin identified Christ Jesus as the principle of universal vitality responsible for all consciousness in primordial cells as well as in complex multicellular organisms:

> Christ... put himself in the position (maintained ever since) to subdue under himself, to purify, to direct and superanimate the general ascent of consciousness in which he inserted himself. By a perennial [i.e., ongoing and everlasting] act of communion and sublimation [i.e., transformation], he aggregates to himself the total *psychism* of the earth. [brackets and italics mine]
>
> *Phenomenon, Epilogue, p. 294*

To express the role of Christ Jesus in subduing and gathering all things unto himself in the process of their ultimate unification and eventual collective presentation to God the Father at the end of *the Millennium,* de Chardin stated:

> And when [Christ Jesus] has gathered everything together and transformed everything, he will close in upon himself and his conquests [through the process of involution], thereby rejoining, in a final gesture, the divine focus he has never left. Then as St. Paul tells us [in *1 Corinthians 15:24-28], God shall be all in all.* [brackets mine]
>
> *Ibid.*

Although the present author sees the formation of the *all in all* (at the end of *the Millennium* of Jesus Christ's rule on Earth) as an infusion of the physically-knowable universe with the Totality of the Creator-God's Being and Fiery Presence, de Chardin's view is somewhat more immanent than transcendent:

> This [formation] is indeed a superior form of 'pantheism' without trace of the poison of adulteration or annihilation: the expectation of perfect unity, steeped in which each element will reach its consummation at the same time as the universe. [brackets mine]
>
> *Ibid.*

Because of his bent toward Aristotelianism — as attested by his embracing "immanence within matter" *(Phenomenon, p. 88)*, de Chardin failed in his writing to provide an account of the Biblical revelation that, when God the Son presents everything *under his feet* (i.e., within his control and power) to God the Father at the time of the formation of the *all in all* referred to in 1 Corinthians 15:28 (KJV), all elements in the physically-observable universe will be consumed by *fervent heat* in a God-induced atomic fission:

> {10} But the day of the Lord will come as a thief in the night, at which time the heavens shall pass away with a great noise, and the elements shall melt with fervent heat, and the earth also and the works that are therein shall be burned up. {11} Seeing then that all these things shall be dissolved, what manner of persons ought you to be in all holy conversation and godliness. {12} Looking for and hastening unto the coming of the day of God, wherein the heavens being on fire shall be dissolved, and the elements shall melt with fervent heat.
>
> *2 Peter 3:10-12 KJV Paraphrase*

For de Chardin, every physical thing has both a "Within" as well as a "Without." The "Without" (i.e., *le Dehors*) comprises a physical thing's external features and characteristics. The "Within" (i.e., *le Dedans*) of a physical thing is consciousness itself, which has urged and pushed physicality to grope toward hominization. To de Chardin, reflection (i.e., self-reflection and knowing that one knows) has played the most important role in the beginning of the hominization of anthropoids. He believed, rightfully so, that when anthropoids crossed the threshold from simply *thinking* to *reflecting*, and from simply *knowing* to *knowing that they knew,* they had evolved into true man (i.e., hominized men and women, or modern hominins). For de Chardin, the psychic advance of prehistoric true man from earlier anthropoids is evidenced by: (1) their increased cerebralization with correspondingly larger cranial cavities; and (2) their behaviors associated with *self-reflection.* The aforementioned behaviors first appeared *as a complete package, or ensemble, of skills* somewhere between 80,000 and 48,000 years ago and included: (1) chipping and polishing stones; (2) making fire in hearths; (3) ritually burying the dead; (4) adorning the living or dead body with scars, inks, tattoos, and/or jewelry; (5) carving and painting on rocks and cave walls; (6) planting crops; (7) making artifacts associated with worship; and (8) producing functional pottery. In his *self-reflection,* the earliest modern man not only recognized his own physical, mental, emotional, spiritual, and social needs but also invented ways to meet those needs.

For the present author, just as *Homo neanderthalensis* and *Homo sapiens* coexisted and interbred in Eurasia for an overlapping 5,000 year period of time (from approximately 44,000 to 39,000 years ago), so also did prehistoric modern man *(Homo sapiens* var. *sine anima)* and the descendants of Adam and Eve *(Homo sapiens* var. *cum anima)*[22] co-exist and interbreed for up to 1656 years (from

[22] See Footnotes 5 and 6 in Volume One for explanations of the present author's nomenclature concerning these two varieties of *Homo sapiens*.

approximately 4004 BC to 2348 BC), which duration is based on the following two criteria: (1) from calculations of Bible genealogies, Adam and Eve materialized on Earth in approximately 4004 BC when they were exiled from the Garden of Eden; and (2) from calculations of successive generations of various Antediluvians in the Bible, specifically: Adam, Seth, Enosh, Kenan, Mahalelel, Jared, Enoch, Methusaleh, Lamech, and Noah (Noah's flood occurring approximately 2348 BC).

Just as *Homo sapiens,* some of whom interbred with *Homo neanderthalensis,* outlasted and replaced *Homo neanderthalensis,* so also did *Homo sapiens* var. *cum anima,* some of whom interbred with *Homo sapiens* var. *sine anima,* outlast and replace *Homo sapiens* var. *sine anima,* resulting in physically-evolved human beings with souls. For the sake of clarity, Adam and Eve materialized in the flesh of hominids (specifically, in the flesh bodies of hominins or modern human beings) and their direct descendants interbred with hominids (specifically, hominins or modern human beings) who had evolved physically but did not have souls. Thus, contemporary members of the species *Homo sapiens* received their physical forms from the descendants of Adam and Eve who interbred — beginning approximately 6,000 years ago — with physically-evolved modern man. Consequently, in keeping with the Will of the Creator-God, all direct descendants of Adam and Eve received eternal souls (albeit fallen eternal souls) as their birthright.

In addition to a geosphere and a biosphere, de Chardin believed that the Earth possesses a surrounding noösphere, or *sphere of human thought,* with an axis of increasing complexity that: (1) drove the evolution of humankind in the direction of reflection as well as socialization and (2) continues to drive its evolution toward higher heights through unification. In other words, de Chardin postulated that the Earth possesses a global human consciousness, or human collective consciousness, that moved humankind from possessing simple instincts to developing traits associated with self-reflection and, then, to developing traits associated with socialization. To de Chardin, traits associated with socialization are of a higher order than traits associated with reflection because they have allowed modern man to reach his current psychic heights and will permit him to reach even higher heights in the future through unification (as human beings converge in the future at *the Omega Point).*

To de Chardin, the future higher heights of human beings include reaching *the Omega Point.* To the present author, the higher heights of humankind along the way to this so-called Omega Point also include human beings developing spiritualized intuition in the form of increased extra-sensory perception, heightened susceptibility to the thoughts of the Creator-God, and enhanced sensitivity to the specific thoughts and feelings of others. The development of this spiritualized intuition is in keeping with the following Bible prophecy:

> {17} "It shall come to pass in the last days," says the Lord God Almighty, "I will pour out My Holy Spirit upon all flesh: and your sons and your daughters shall prophesy, and your young men shall see visions, and your old men shall dream dreams; {18} And on my servants and on my handmaidens in those days I will pour out My Holy Spirit; and they shall prophesy."
>
> *Acts 2:17-18 KJV Paraphrase (quoting Joel 2:28-29)*

To be sure, the Creator-God has not chosen any human being to be the sole voice of truth for this "psychozoic era" *(Phenomenon, p. 183).* Rather, the Creator-God has chosen His Holy Spirit to be the sole voice of truth for this age. How will humankind know when its psychism has arrived at its highest heights? For the present author, we shall arrive — as well as recognize that we have arrived — when our thoughts in relative space-time are no longer distinguishable from thoughts in eternity.

Like de Chardin, the present author believes that there is a role for our enemies to play in the development of higher psychisms in modern man. (For the present author, higher psychisms include all mental activities associated with *spiritualized intuition*.) So important is the role of this impetus for change, de Chardin asked "What would we do without our enemies?" *(Phenomenon, p. 149)* In other words, our enemies provide stimuli for us to reach higher psychisms by helping us to move away from our "fundamental inertia" *(Ibid.)* — in conjunction, of course, with the outpouring of the Creator-God's Holy Spirit, such outpouring first referred to by the Prophet Joel *(Joel 2:28-29)*. Our enemies play an important role in our arriving at higher heights because their potential, imminent, and actual intrusions or invasions force the development of our spiritualized intuition to help ensure our individual and collective survival.

The present author believes that the higher psychisms of increased extra-sensory perception, heightened susceptibility, and enhanced sensitivity are examples of spiritualized intuition that accompany the higher consciousness provided to us by our Creator-God through His Holy Spirit. In contrast to de Chardin's requirement of an innate "*inner* principle" *(Phenomenon, p. 149)* for increased psychogenesis, the present author views the human brain of modern man more as a channel, or conveyor, for heightened consciousness rather than as an originator of increased consciousness. As once told to me by a heavenly source: "The family of God increases by decreasing, includes by excluding, and often varies yet never changes." Metaphysically speaking, the family of God excludes its enemies at the same time that it sharpens its wits in spiritualized intuition through the Creator-God's Holy Spirit in order to protect itself from potential intrusions as well as imminent invasions. Christ Jesus forewarned his disciples: "Understand that I send you out as sheep in the midst of wolves: therefore, be wise as serpents and harmless as doves" *(Matthew 10:16 KJV Paraphrase)*. Spiritualized intuition mediated by the Creator-God's Holy Spirit is an integral part of being wise and providing an effective witness in addition to surviving.

Much ado has been made about de Chardin's *Omega Point* by all sorts of people who want to make the concept more complicated than it is. To de Chardin, evolution is simply the rise of consciousness, and the rise of consciousness eventually effects a psychic union of all human beings, whose psychic union is named by de Chardin as *the Omega Point.* For de Chardin, "no evolutionary future awaits man except in association with all other men" *(Phenomenon, p. 246)*. In the elaboration of his *Omega Point,* de Chardin placed hope in mankind by developing his own peculiar brand of religious humanism; to be sure, he believed that "the crown of [human] evolution" is situated "in a supreme act of collective vision obtained by a pan-human effort of investigation and construction" *(Ibid., p. 249, brackets mine)*. Unfortunately, de Chardin's ideal human government consists of an elite intelligentsia dominating the masses *(Ibid., Footnote 1, p. 245)*. Although de Chardin decried the injustices of Communism and National Socialism, he might not have condemned the most modern form of totalitarianism (i.e., the Beast of Islam under the control of the final, end-time Antichrist). To be sure, de Chardin's fascination with populist unanimity and mechanization is seen in his comment about perverted idealism:

> Monstrous as it is, is not modern totalitarianism really the distortion of something magnificent, and thus quite near to the truth?
>
> *Phenomenon, p. 256*

The previous quote from de Chardin is equivalent to saying that Stalin and Hitler were geniuses. Although one can argue for the genius status of those despots, it is not expedient to do so because their cunning was a perversion of the Creator-God's nature of intelligence that He intended for humankind.

For those who might erroneously assume that de Chardin's *Omega Point* is equivalent to *Christ Jesus, Christ Consciousness,* or a *Cosmic Christ,* de Chardin provides clarification that there is still "a supreme Someone" who supersedes his human-based *Omega Point:*

> To be more exact, "to confirm the presence at the summit of the world of something in line with, but still more elevated than, the Omega point." This is in deference to the theological concept of the "supernatural" according to which the binding contact between God and the world, *hic et nunc* inchoate [here and now not fully formed], attains to a super-intimacy (hence also a super-gratuitousness) of which man can have no inkling and which he can lay no claim by virtue of his "nature" alone. [brackets mine]
>
> *Phenomena, Epilogue, Footnote 1, p. 298*

Thus, Christ Jesus himself is "already on high" *(Phenomenon, p. 298)* — at a summit far beyond the locus in which a unified and evolved human consciousness can, and will, converge — which locus is called *the Omega Point* by de Chardin. To be sure, without reading one footnote in its Epilogue (see previous quote), the student of de Chardin's *The Phenomenon of Man* might dispute the existence of the actual truth to which de Chardin subscribed (i.e., that there is something beyond *the Omega Point).*

2.5.4.3 The Metaphysics of de Chardin in *The Phenomenon of Man*

Either de Chardin was not entirely honest with himself or he was not entirely honest with his readers, reviewers, and colleagues. (Certainly, both types of dishonesty can be true at the same time as well.) And, because we cannot determine it conclusively, either his dishonesty was intentional or it was unintentional. The present author has made his own conclusions based on what de Chardin himself has written in *The Phenomenon of Man:*

Although sufficient definitions and ample explanations and examples of *metaphysics* have been provided in Volume One of

Intelligent Evolution, the simplest and best definition of *metaphysics* (i.e., the one that the present author believes has the greatest utility for the readers of *Intelligent Evolution*) is "the abstract study of invisible reality as it relates to the cause and essence of physical reality and its purpose for existence." Applying this definition as well as a number of other accepted alternative definitions, de Chardin's *The Phenomenon of Man* is definitely a metaphysical work. However, de Chardin issued this disclaimer in the Preface to his book:

> If this book is to be properly understood, it must be read not as a work on metaphysics, still less as a sort of theological essay, but purely and simply as a scientific treatise. The title itself indicates that. This book deals with man solely as a phenomenon; but it also deals with the whole phenomenon of man.
>
> *Phenomenon, Preface, p. 29*

Regardless of the direction in which de Chardin wanted to steer his readers, the present author believes that *The Phenomenon of Man* must be read as a work on metaphysics because that is what it is *in toto.* It is not *simply* "a scientific treatise." To be sure, there are many scientific facts presented in de Chardin's book from the areas of anthropology, archeology, biochemistry, biology, chemistry, cosmology, geology, paleontology, and physics, but there are many metaphysical concepts presented in it as well — some blatantly metaphysical and others implied. And the overarching theme of *The Phenomenon of Man* that provides coherence to that work is metaphysical in essence. The present author believes that it would have been more accurate for de Chardin to title his book *The Causality and Phenomenology of Man.*

To be sure, although de Chardin's work addressed the phenomenology of man, it also addressed the noumenology of man (even though de Chardin did not care to openly admit that it did). As an academically-trained and credible biologist, geologist, naturalist, and paleontologist with extensive professional experience in those areas, perhaps de Chardin intentionally made this blunder to redirect the reader's attention away from his philosophical and theological views on cosmology and ontology. Perhaps, from de Chardin's perspective, there would be less criticism of his work if readers and reviewers looked at it as a scientific rather than a metaphysical treatise. Whatever the reason, it is considered a blunder by the present author because the work would have stood the test of time better if de Chardin had simply labeled *The Phenomenon of Man* a metaphysical work with underpinnings in anthropology, archeology, biochemistry, biology, chemistry, cosmology, geology, paleontology, and physics.

In what way would de Chardin's work have stood the test of time better as a metaphysical rather than a scientific treatise?

For the present author, calling a written work metaphysical when it is metaphysical helps to remove it from the fray of unwarranted criticisms concerning its scientific nature because readers and reviewers will look at it as speculative (as they should look at it). Except, perhaps, for unintended inaccuracies about scientific facts or for clearly irrational argumentation and/or magical thinking, true metaphysical works are lifted an extra degree above criticism because they are representative of the abstract thinking and speculation of (1) one thinker or (2) a group of thinkers who think similarly. Metaphysics is all about thinking. Who can discredit thinking when it is meant to be abstract and speculative? Certainly, one can reject the conclusions of a particular thinker, but one cannot say that the work of a particular thinker does not deserve an objective hearing. *For example,* one may conclude that *Intelligent Evolution* is fanciful but one cannot conclude that it should never have been written or published. Why not? It is the present author's right to do both regardless of who agrees or does not agree with the concepts he has presented in his work.

There is a great deal of relevance in this statement by Lewis Carroll from *Through the Looking-Glass (1872, Chapter Six, p. 205):*

"'When I use a word,' Humpty Dumpty said, in rather a scornful tone, 'it means just what I choose it to mean — neither more nor less.'" However, just because de Chardin chose to deny the label of *metaphysics* for *The Phenomenon of Man* does not make it so. Rather, what de Chardin actually wrote throughout his book makes it so or does not make it so. And what de Chardin has written throughout *The Phenomenon of Man* makes it undeniably *metaphysical* rather than *not metaphysical.*

In Section 2.5.3 of *Intelligent Evolution,* the present author stated that "just because language is metaphorical does not make it metaphysical and, conversely, just because language is metaphysical does not make it metaphorical." Relative to metaphysical language and written imagery employing simile, metaphor, or personification, de Chardin's *The Phenomenon of Man* has three categories: (a) writing that uses imagery but is not genuinely metaphysical; (b) writing that uses imagery and is genuinely metaphysical; and (c) writing that is genuinely metaphysical but does not use imagery. Representative examples of these three categories in *The Phenomenon of Man* follow:

(a) The Use of Imagery that is not Genuinely Metaphysical

The following quotes from *The Phenomenon of Man* illustrate that, when language becomes too flowery and/or overly personified, it loses the metaphysical value it might have had:

> Refracted rearwards along the course of evolution, consciousness displays itself qualitatively as a spectrum of shifting shades whose lower terms are lost in the night.
>
> *Phenomenon, p. 60*

> Let us have a look at the earth in its early stages, so fresh yet charged with latent powers, as it balances in the chasms of the past.
>
> *Ibid., p. 67*

A circle can augment its order of symmetry and become a sphere.

Ibid., p. 89

The elemental ripple of life that emerges from each individual unit does not spread outwards in a monotonous circle formed of individual units exactly like itself. It is diffracted and becomes iridescent, with an indefinite scale of variegated tonalities.

Ibid., p. 105

And then, so it seems, so as to enlarge the breach thus made by its first inroads in the ramparts of the unorganised world, life discovered the wonderful process of conjugation.

Ibid., p. 106

Sometimes the new subdivisions seem merely to correspond to superficial diversifications — they are effects of chance or of a playful inventive exuberance.

Ibid., p. 117

... the elements of a phylum tend to come together and form societies just as surely as the atoms of a solid body tend to crystallise.

Ibid., p. 118

In the course of this struggle to master the dimensions and the relief of the universe, space was the first to yield — naturally, because it was more tangible. In fact the first hurdle was taken in this field when long, long ago a man (some Greek, no doubt, before Aristotle), bending back on itself the apparent flatness of things, had an intuition that there were antipodes. From then onwards round the round earth the firmament itself rolled roundly.

Ibid., p. 217

In order for language that employs imagery to be useful either to pure science or to pure metaphysics, it needs to be written, or rewritten, without attempts to be cute, flowery, or forced. Indeed, today's writers of pure science will write, or rewrite, evolution-related statements to reflect no anthropocentric or theocentric sentiments and no purpose for any evolutionary change even if the change is advantageous for the survival of an organism or a group of organisms. And, in contrast, today's writers of pure metaphysics would write, or rewrite, evolution-related statements to reflect anthropocentric and/or theocentric sentiments as well as purpose-related reasons for each evolutionary change:

For example, "life discovered the wonderful process of conjugation" *(Phenomenon, p. 106)* could be rewritten in the language of pure science to make clear that living things do not consciously incorporate into their populations processes that are advantageous to their survival; rather, groups of living things (i.e., populations, species, or phyla) incorporate changes that have occurred by chance that then prove themselves to be either advantageous or disadvantageous for survival. But the same statement "life discovered the wonderful process of conjugation" *(Ibid.)* could be rewritten in the language of pure metaphysics to make clear that, although living things do not consciously incorporate into their populations processes that are advantageous to their survival, an immanent or transcendent force *directs* (not necessarily "discovers" as de Chardin would allege) any and all changes to ensure the emergence, survivability, sustainability, and thrivability of a specific population, species, or phylum — *for instance*, of *Homo sapiens* for the ultimate purpose of permitting salvation opportunities for fallen souls who reside in human bodies.

(b) The Use of Imagery that is Genuinely Metaphysical

Before proceeding with this section, it is important for the reader to understand that the present author herewith gives examples of de Chardin's use of imagery that is genuinely metaphysical but with which the present author may or may not agree concerning the nature of its metaphysical truth:

> The consciousness of each of us is evolution looking at itself and reflecting upon itself. With that very simple view... a new light — inexhaustibly harmonious — bursts upon the world, radiating from ourselves.
>
> *Phenomenon, p. 221*

> Either nature is closed to our demands for futurity, in which case thought, the fruit of millions of years of effort, is stifled, still-born in a self-abortive and absurd universe. Or else an opening exists — that of the super-soul above our souls; but in that case the way out, if we are to agree to embark on it, must open out freely onto limitless psychic spaces in a universe to which we can unhesitatingly entrust ourselves.
>
> *Ibid., p. 233*

> The coalescence of elements and the coalescence of stems, the spherical geometry of the earth and psychical curvature of the mind harmonising to counterbalance the individual and collective forces of dispersion in the world and to impose unification — there at last we find the spring and secret of hominisation.
>
> *Ibid., p. 243*

> Now from this point of view and in the present condition of things, there are two ways, through two stages, in which we can picture the form mankind will assume tomorrow — either (and this is simpler) as a common power and act of knowing and doing, or (and this goes much deeper) as an organic superaggregation of souls.
>
> *Ibid., p. 248*

Metaphysical language that uses imagery (i.e., simile, metaphor, or personification) abounds throughout *The Phenomenon of Man.*

For example, de Chardin used personification when he stated: (1) that nature has "groped" and will continue to "grope" to find new ways to "invent" itself; and (2) that consciousness from its outset has been "groping" for its consummation through physical evolution and its continuing ascent in biological complexity (see also *Phenomenon of Man, pp. 237-238):*

> *Groping* is directed chance. It means pervading everything so as to try everything, and trying everything so as to find everything. Surely in the last resort it is precisely to develop this procedure (always increasing in size and cost in proportion as it spreads) that nature has had recourse to profusion. [italics mine]
>
> *Phenomenon, p. 110*

> It was a marvellous period of investigation and *invention* when, in the unequalled freshness of a new beginning, the eternal *groping* of life burst out in conscious reflection. [italics mine]
>
> *Ibid., p. 205*

Concerning his use of personification, de Chardin admitted that the words "groping" and "invention" are both "imbued... with anthropomorphism" *(Phenomenon, p. 223)*. And, concerning de Chardin's metaphysics, the present author acknowledges that the consciousness behind de Chardin's evolution is either not intelligent or not as intelligent as it should be if de Chardin is referring to the Supraconsciousness of the Creator-God, the One-who-knows all outcomes before He causes them to come to pass.

(c) Genuinely Metaphysical Language that does not use Imagery

To de Chardin, *consciousness* was, and is, the single unifying and most coherent factor that binds the physical universe together. However, rather than attribute all consciousness to the immanent and/or transcendent Mind of the Creator-God, de Chardin assumed the perspective that "consciousness reveals itself as a cosmic property of variable size subject to a global transformation" *(Phenomenon, p. 58).* To be sure, the immanence of consciousness is evident in de Chardin's particular phenomenology: He believed that consciousness was, and is, always present but that its presence was not really evident until an individual knower could know himself or herself in thought through reflection, culminating in the knower's understanding that biological evolution manifested itself in phylogenesis as the ramifications of a biological tree of life. (For the sake of clarity, a *phylum* is a group of organisms that ranks above class and below kingdom, and *phylogenesis* is the evolutionary development and diversification of a group of organisms either into a population within one phylum or into an entirely new phylum.)

Indeed, de Chardin concluded that simple things possess less consciousness and that complex things possess more consciousness:

> The degree of concentration of a consciousness varies in inverse ratio to the simplicity of the material compound lined by it. Or again: a consciousness is that much more perfected according as it lines a richer and better organised material edifice.
>
> *Phenomenon, p. 60*

> Spiritual perfection (or conscious 'centreity') and material synthesis (or complexity) are but the two aspects or connected parts of one and the same phenomenon.
>
> *Ibid., pp. 60-61*

> To connect the two energies, of the body and the soul, in a coherent manner: science has provisionally decided to ignore the question, and it would be very convenient for us to do the same. Unfortunately, or fortunately, caught up as we are here in the logic of a system where the *within* of

> things [de Chardin's *le Dedan des Choses*] has just as much or even more value than their *without* [de Chardin's *le Dehors*], we collide with the difficulty head on. It is impossible to avoid the clash: we must advance. [italics and brackets mine]
>
> *Ibid., p. 62*

The two previous quotes beg this question: "Does a purely *scientific treatise* ever: (1) discuss *spiritual perfection* except to point the reader to someone else's written work about it; or (2) try to connect the energies of the physical body [i.e., the Without of Things, or *le Dehors des Choses*] and the spiritual soul [i.e., the Within of Things, or *le Dedan des Choses*]?" The unabashed answer is "No, it does not." A purely scientific treatise would never discuss spiritual perfection except to state that the topic is outside of the realm of natural science. And a purely scientific treatise would never discuss *the Within of Things* except in a physically-elemental, -atomic, or -subatomic sense when using descriptive imagery — and, certainly, never in a noumenal sense.

De Chardin believed that "a certain mass of elementary consciousness was originally emprisoned in the matter of earth" *(Phenomenon, p. 72)* and this imprisoned consciousness gave rise to the earliest forms of life on the planet Earth. Although it provides a metaphysical explanation for the origin of life, de Chardin's conclusion is inconsistent with the paradigm of *intelligent evolution* put forth by the present author, in which all direction to physical evolution is given by the transcendent Creator-God and all outcomes are pre-determined as well as foreknown by Him.

In the final analysis, de Chardin's *The Phenomenon of Man* is a metaphysical treatise with scientific underpinnings. It is not a scientific treatise. (Hopefully, the readers of *Intelligent Evolution* will not consider the last two sentences negative criticisms of de Chardin's work.)

2.5.4.4 The Relevance of Lamarck and Haeckel

Both Jean-Baptiste Pierre Antoine de Monet, Chevalier de Lamarck (1744-1829) and Ernst Heinrich Philipp August Haeckel (1834-1919) are important to the theory of evolution in general and to the paradigm of *intelligent evolution* in particular. They are included in this section on de Chardin because de Chardin referred to each of them: He referred to Lamarck by name as well as in discussing Lamarck's work, and, without specifically naming Haeckel, de Chardin clearly utilized Haeckel's theory — that *ontogeny recapitulates phylogeny* — many times throughout *The Phenomenon of Man*. As a side note, it is important for the readers of, or listeners to, de Chardin's work not to confuse the surname Haeckel with the surname Hegel, whom de Chardin also referred to in *The Phenomenon of Man*. Georg Wilhelm Frederich Hegel (1770-1831) was a philosopher generally considered responsible for German idealism and from whom Eddy was accused, incorrectly so, of purloining her metaphysical thinking.

Lamarck

In his book, *Zoological Philosophy: An Exposition with Regard to the Natural History of Animals,* published in French in 1808, Lamarck specifically refers to himself as a "naturalist and physicist" *(p. 184)*. However, for the modern reader, it would be more accurate to state that Lamarck was a French botanist, zoologist, and taxonomist. And, for the purpose of clarification at this juncture, a *taxonomist* is someone: (1) who categorizes organisms into groups based on their physical and physiological characteristics as well as their probable evolutionary relationships; and (2) who names species based on accepted rules for such naming (i.e., newly-discovered species, species that have not been named previously, or species that have not been named correctly).

Lamarck's *Zoological Philosophy* is an eloquently written book, and its 1914 English translation by Hugh Elliot is presented in beautifully flowing, well-written, and easy-to-understand language. *Zoological Philosophy* could easily be used as one of the textbooks for a course in the history of natural philosophy/natural science (see Section 2.4 — entitled *Proposed Curriculum for the Millennium*).

In his *Zoological Philosophy,* Lamarck presented the following practical definitions for *species* and *nature:*

> Any collection of like individuals which were produced by others similar to themselves is called a species.
>
> *Ibid., p. 35*

> Nature... cannot be for us more than the totality of objects comprising: (1) all existing physical bodies; (2) the general and special laws, which regulate the changes of state and position to which these bodies are liable; (3) lastly, the movement distributed at large among them, which is continually preserved or being renewed, has infinitely varied effects, and gives rise to that wonderful order of things which this totality embodies.
>
> *Ibid., p. 183*

Lamarck believed that there was a solid basis for his articulating a philosophy of zoology:

> It is known that every science must have its philosophy, and that it cannot make real progress in any other way. It is in vain that naturalists fill their time in describing new species, in grasping all the shades and small details of their varieties, in enlarging the immense list of catalogued species, in establishing genera, and in making incessant changes in the principles which they use. If the philosophy of science is neglected [scientific] progress will be unreal, and the entire work will remain imperfect. [brackets mine]
>
> *Ibid., p. 33*

For Lamarck, the best *zoological philosophy* was embodied in the scientific theory of evolution that challenged the then-held notion of the *fixity* of all species (i.e., the unchanging nature of individual groups of organisms). Nevertheless, Lamarck was not a godless evolutionist: Although Lamarck restricted himself to the role of "a pure observer of nature," he wrote the following about our Creator-God:

> Doubtless, nothing exists but by the will of the Sublime Author of all things, but can we set rules for him in the execution of his will, or fix the routine for him to observe? Could not his infinite power create an order of things which gave existence successively to all that we see as well as to all that exists but that we do not see? Assuredly, whatever his will may have been, the immensity of his power is always the same, and in whatever manner that supreme will may have asserted itself, nothing can diminish its grandeur. I shall then respect the decrees of that infinite wisdom and confine myself to the sphere of a pure observer of nature. If I succeed in unravelling anything in [nature's] methods, I shall say without fear of error that it has pleased the Author of nature to endow [nature] with that faculty and power. [brackets mine]
>
> *Ibid., p. 36*

> I prefer to think that the whole of nature is only an effect: hence, I imagine and like to believe in a First Cause or, in short, a Supreme Power which brought nature into existence and made it such as it is.
>
> *Ibid., pp. 183-184*

To be sure, there is no specific evidence that Lamarck knew Christ Jesus as his personal Lord and Savior, but he certainly was headed in the right direction — and pointed others in that direction as well. In agreement with the Biblical account of the sequential creation of new groups of organisms, Lamarck wrote:

> If indeed it is true that all living bodies are productions of nature, we are driven to the belief that [nature] can only have produced them one after another and not all in a moment. Now if [nature] shaped them one after another, there are grounds for thinking that [nature] began exclusively with the simplest, and only produced at the very end the most complex organisations both of the animal and [plant] kingdoms. [brackets mine]
>
> *Ibid., p. 129*

In his *Zoological Philosophy,* Lamarck laid out the first cohesive and coherent theory of evolution based on: (1) the power of life, a force that drives complexification in evolution; and (2) the influence of circumstances, such influence an adaptive force that operates through the use and disuse of body parts, regions, and organs. Lamarck's theory posited that physical and physiological changes from the use or disuse of body parts, regions, and organs were passed on from one generation to the next. Through his theory, a case was made for the malleable nature of general characteristics of individual groups of organisms.

Concerning Lamarck's theory of evolution, he articulated these two laws in his *Zoological Philosophy:*

> *First Law.* In every animal which has not passed the limit of its development, a more frequent and continuous use of any organ gradually strengthens, develops and enlarges that organ, and gives it a power proportional to the length of time it has been so used; while the permanent disuse of any organ imperceptibly weakens and deteriorates it, and progressively diminishes its functional capacity, until it finally disappears.
>
> *Second Law.* All the acquisitions or losses wrought by nature on individuals, through the influence of the environment in which their race has long been placed, and hence through the influence of the predominant use or permanent disuse of

> any organ; all these are preserved by reproduction to the new individuals which arise, provided that the acquired modifications are common to both sexes, or at least to the individuals which produce the young.
>
> *Ibid., p. 129*

Lamarck acknowledged that: (1) use or disuse was not the only factor that caused changes in groups of organisms; and (2) not all organs or parts — *for example,* those associated with hearing — are changed through their use or disuse:

> If the factor which is incessantly working towards complicating organisation were the only one which had any influence on the shape and organs of animals, the growing complexity of organisation would everywhere be very regular. But it is not; nature is forced to submit her works to the influence of their environment, and this environment everywhere produces variations in them. This is the special factor which occasionally produces in the course of the degradation that we are about to exemplify, the often curious deviations that may be observed in the progression.
>
> *Ibid., p. 69*

> Progress in complexity of organisation exhibits anomalies here and there in the general series of animals, due to the influence of environment and of acquired habits.
>
> *Ibid., p. 70*

Examples of changes in groups of organisms that Lamarck used to support his claims included organs and parts associated with the visual system that did not develop because of their lack of exposure to light in specific cave-dwelling species. Not only did the organs and parts not develop in cave environments but also did they not develop in the offspring of those species when their parents had been removed from the caves and placed in environments with daily exposure to sunlight. To be sure, contemporary evolutionary theory would add that, by chance, there had been natural variations in the cave-dwelling species that included those with the capacity to form normal photoreceptors if light were present during early development as well as those with underdevelopment of the visual system, and that survival of progeny with underdeveloped visual systems was favored for multiple reasons related to complex environmental and inheritance factors.

To reiterate what the present author has stated previously, contemporary evolutionary theory does not have an individual organism consciously adapting to its environment but, rather, a group of organisms demonstrating an adaption based on: (1) natural variations within its population, (2) continued survival of the organisms within the group because of the existence of advantageous or non-disadvantageous traits or characteristics, and (3) inheritance of factors associated with those advantageous or non-disadvantageous traits or characteristics — until all, or at least a majority, of the members of the group under consideration possess a specific trait or characteristic.

Related to Lamarckism, Teilhard de Chardin understood that he might be accused of being too Lamarckian by attributing evolutionary changes to the *groping, inventing, discovering, urging, pushing,* and *grasping* of consciousness. However, de Chardin also tried to show the complementary aspects of Lamarckism and Darwinism by stating:

> In various quarters I shall be accused of showing too Lamarckian a bent in the explanations which follow, of giving an exaggerated influence to *the Within [au Dedans]* in the organic arrangement of bodies. But be pleased to remember that, in the 'morphogenetic' action of instinct as here understood, an essential part is left to the Darwinian play of external forces and to chance. It is only really through strokes of chance that life proceeds, but strokes of chance which are recognised and grasped — that is to say,

> psychically selected [i.e., chosen by consciousness]. Properly understood the 'anti-chance' of the Neo-Lamarckian is not the mere negation of Darwinian chance. On the contrary it appears as its utilisation. There is a functional complementariness between the two factors; we could call it 'symbiosis'. [italics and brackets mine]
>
> *Phenomenon, Footnote 1, p. 149*

When de Chardin wrote *The Phenomenon of Man,* Lamarckism as well as Neo-Lamarckism had fallen largely into disfavor with the majority of evolutionary biologists. However, since that time, some scientists have recognized the possibility of behavior suppressing or repressing the operation of certain genes within an organism without changing the genetic sequence in the particular code for that organism yet changing its traits or characteristics. This field of study is referred to today as *epigenetics.*

Reading in tandem de Chardin's *The Phenomenon of Man* and Lamarck's *Zoological Philosophy* has permitted the present author to make the following important and relevant conclusions:

1. Neo-Lamarckism and Neo-Darwinism are not only complementary but also capable together of providing major scientific insights into the processes of evolution that incorporate principles of use, disuse, natural variation, survival of the fittest through natural selection, transmission of traits and characteristics from one generation to the next and resulting group adaptation to a changing environment.

2. A godless theory of evolution is the product of an imagination that has many related and seemingly unrelated facts available to it but whose author *does not* have a personal relationship with the God of the Holy Bible; in contrast, a godly theory of *intelligent evolution* is the product of an imagination that has many related and seemingly unrelated facts available to it and whose author *does* have a personal relationship with the God of the Holy Bible — in

particular through His *only-begotten* Son, Christ Jesus, and through His Holy Spirit.

3. Animals with the highest form of intelligence know by reflection that they exist; animals with the highest form of intelligence are also capable of knowing that their Creator exists but only if they are nurtured, educated, and trained in emotionally-, mentally-, and spiritually-healthy ways; advanced cerebralization in an animal permits the exercise of that animal's individual will; and the level of complexity in cerebralization evidenced in modern *Homo sapiens* allows for the inhabitance of the human body by a fallen eternal soul who, through Christ Jesus, is enabled to once again yield his or her will to the Will of the Creator-God. (Here, *once again* refers to the return of a soul to the original relationship it had with the Creator-God before the Adamic Fall.)

Haeckel

Ernst Haeckel (1834-1919) was a German biologist, naturalist, and taxonomist who was also a medical doctor, comparative anatomist, comparative embryologist, and gifted artist. Although Haeckel promoted Darwinism, he also held to many tenets of natural philosophy developed by Lamarck and Goethe. Haeckel is attributed with helping to popularize the now commonly-used word *phylum,* which refers to a group of organisms with general characteristics below the taxonomic level of *kingdom* but above the taxonomic level of *class* (notwithstanding its rank in relationship to the derived terms *superphylum* and *subphylum*).

A basic understanding of the following biological terminology is essential for the readers of *Intelligent Evolution* in order for them to understand the biogenetic law, or recapitulation theory, for which Haeckel is most well-known: *embryology, embryogenesis, morphology, morphogenesis, ontogeny, ontogenesis, organ, organogenesis, phylogeny, phylogenesis,* and *vertebrate* as well as *invertebrate.*

Haeckel's biogenetic law, or recapitulation theory, is best expressed by the dictum *ontogeny recapitulates phylogeny*. For the overwhelming majority of people familiar with this dictum (including the overwhelming majority of biologists), it is only a sound bite representing a theory that is fraught with scientific and artistic inaccuracies and political, as well as professional, incorrectness. Such people react to Haeckel's dictum as if it was promoted and embraced as a dogma rather than a flexible principle of parallel embryogenesis nicely and neatly explained by its author with many qualifications. In other words, most people (even most biologists) have never read the conceptual guidelines provided by Haeckel himself, which guidelines include the following:

> To ensure accuracy, we must first become acquainted with the various mental operations which we shall apply in this natural-philosophical research. These operations are partly of an inductive, partly of a deductive nature: partly conclusions from numerous particular experiences to a general law; partly conclusions from this general law back to particular experiences.
>
> *The Evolution of Man, Vol. 2, p. 35*

> There can be no doubt that (if the theory of descent is correct) Man [humankind] has developed as a true Vertebrate, and that he originated from one and the same common parent-form with all other Vertebrates. This special deduction must be regarded as quite certain, correctness of the inductive law of the theory of descent being of course first granted. [brackets mine]
>
> *Ibid., p. 37*

> We can, moreover, name a series of different forms of the vertebrate tribe, which may be safely regarded as the *representatives* of different successive phylogenetic stages of evolution, or as different members of the human ancestral-line. [italics mine]
>
> *Ibid.*

> ...the special proof of all separate parent-forms must always remain more or less incomplete and hypothetical. This is quite natural. For all the records of creation upon which we rely are in a great measure incomplete, and will always remain incomplete; just as in the case of Comparative Philology.
>
> *Ibid., p. 38*

> ...the palaeontological record of creation... will always remain extremely incomplete. Not less incomplete is the second most important record of creation, that of Ontogeny. For the Phylogeny of the individual it is the most important of all. Yet, it also has its great defects, and often leaves us in the lurch.
>
> *Ibid., p. 39*

> The reproduction of the Phylogeny in the Ontogeny is but rarely tolerably complete.
>
> *Ibid.*

> ...in the Ontogeny of the higher animal forms, the Phylogeny has been very greatly limited by Kenogenesis [i.e., the introduction during embryogenesis of structures not found in lower, earlier, or ancestral groups of organisms]; as a rule, only a blurred and much vitiated [i.e., spoiled] picture of the original course of evolution of their ancestors now lies before us in the Ontogeny. [brackets mine]
>
> *Ibid., p. 40*

Concerning his written works, Haeckel never articulated his theory of recapitulation in an unwavering, dogmatic way. Instead, he detailed his theory of recapitulation with many well thought out qualifications within the context of using inductive and deductive reasoning. Perhaps the only legitimate criticisms concerning Haeckel's works have to do with his artistic renderings of certain vertebrate embryos.

Substantive criticisms of Haeckel's artistic renderings of certain vertebrate development (represented in Figure Three of *Intelligent Evolution*) include that:

1. The embryos of the individual organisms were not drawn to scale in Haeckel's artistic renderings. In other words, different organisms were represented as the same size. However, Haeckel acknowledged in his *Anthropogenie* that he did this purposely to facilitate comparison of the embryos.

2. Certain embryonic structures were omitted in Haeckel's artistic renderings — *for example,* the amnion, allantois, and yolk sac. However, Haeckel acknowledged in his *Anthropogenie* that he purposely omitted them to facilitate comparison of the embryos.

3. Certain additional structures were omitted or underrepresented in Haeckel's artistic renderings, and their omissions or underrepresentations were not explained by Haeckel.

4. Certain additional structures were exaggerated in Haeckel's artistic renderings, and their exaggerations were not explained by Haeckel.

Regardless of any and all criticisms of Haeckel's artistic renderings of vertebrate development made in the mid- to late-nineteenth century, the following three principles are stated by the present author with certainty based not only on Haeckel's works but also on the works of current comparative embryologists:

1. Using a stereoscopic (dissecting) microscope and a photomicroscope with their given limits of resolution, the earliest stages of cleavage and germ layer formation in whole and sectioned *vertebrates* have similarities to the earliest stages of cleavage and germ layer formation in whole and sectioned *invertebrates* (stages for a representative invertebrate are shown in Figure Two of *Intelligent Evolution*).

2. Using a photomicroscope with its given limits of resolution, whole embryos from the seven different vertebrate classes (see Footnote 28 in *Intelligent Evolution*) cannot be easily distinguished from one another at their earliest stages of development (i.e., up through initial formation of neural folds and somites), especially if the embryonic tissues have been stained and if the embryos have had all of their associated extraembryonic structures removed.

3. Using a stereoscopic (dissecting) microscope with its given limits of resolution, embryos from the seven different vertebrate classes (see Footnote 28 in *Intelligent Evolution*) cannot be easily distinguished from one another at their earliest stages of development (i.e., up through the initiation of pharyngeal arches, pharyngeal pouches, and pharyngeal clefts or slits), especially if the embryos have had all of their associated extraembryonic structures removed.

If the present author had been required to teach biology in an environment in which he was forbidden to mention the word, concept, paradigm, or theory of evolution or mention the word, concept, paradigm, or theory of recapitulation, then he would have taught comparative anatomy, comparative embryology, homology of vertebrate structures, and analogy of vertebrate structures and allowed the more intelligent and imaginative students to draw (i.e., make) their own inferences. Similarly, if the present author had pastored a church congregation in a country where the government expressly outlawed speaking out against Islam, he would have read aloud passages from the Bible and contradictory passages from the Qur'an and allowed each member of the congregation to make his or her own conclusions without making related conclusions for them. (Of course, the present author would always openly confess Christ Jesus as his personal Savior and the *only-begotten* Son of the God of the Holy Bible regardless of circumstance or consequence.)

For the purpose of historical reference, two of Haeckel's drawings are represented on the following two pages. Figure Two shows the earliest cleavage and germ layer stages of a representative invertebrate (a coral). Figure Three shows embryonic and fetal development from four different mammalian taxonomic orders.

Concerning the embryos of individual organisms not being drawn to scale in Haeckel's artistic renderings in Figure Three, the present author would make use of adult skeletal models of representative vertebrates in a similar way in order to teach: (1) their comparative anatomy; and (2) homology of their bony elements. If cost was not a factor, the present author would make use of five foot (1.524 meter) models of representative vertebrates. Instead of the hall of mirrors at the sumptuous palace in Versailles, the present author envisions a grand gallery of vertebrate skeletons in a learning hall of natural history. (Instead of a sun-god representation of Louis XIV on its ceiling, there would be a representation of Christ Jesus to acknowledge him as the Creator-Evolver of all biological life.)

As indicated previously, de Chardin did not mention Haeckel by name, but de Chardin did utilize Haeckel's principle of recapitulation several times throughout *The Phenomenon of Man:*

> ... everything points to their [forms] being representative forms, even if only as a surviving residue of some particular stage in the construction of terrestrial matter [inanimate or animate]. [brackets mine]
>
> *Phenomenon, p. 84*

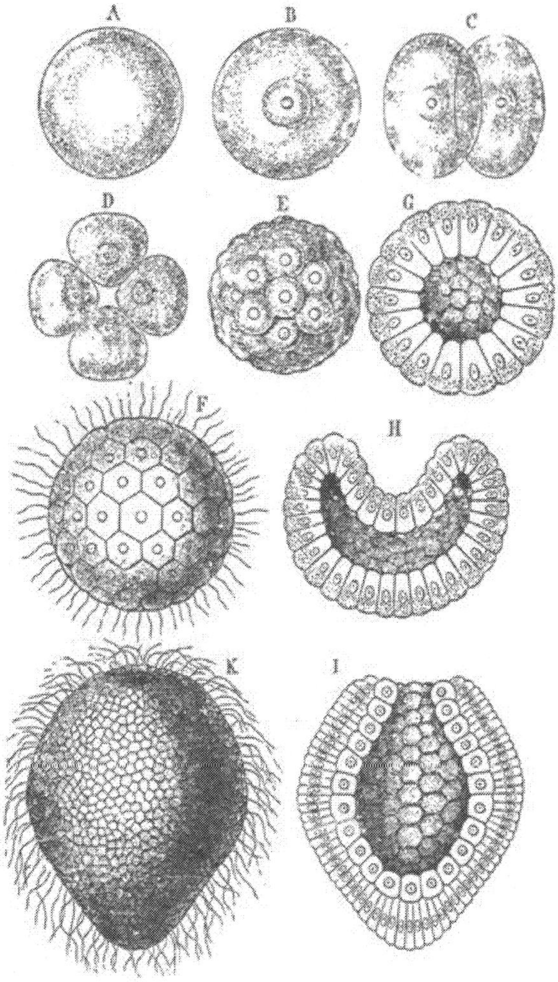

Woodcut Drawing, Figure 171 from Haeckel's 1897 Edition of *Evolution of Man: Volume Two* (page 57)

SHOWING GERMINATION STAGES OF A CORAL AT ONE-CELL (A,B), TWO-CELL (C), FOUR-CELL (D), MORULA (E), BLASTULA CROSS SECTION (G), BLASTULA WHOLE (F), EARLY GASTRULATION (H), GASTRULA LONGITUDINAL SECTION (I), AND GASTRULA WHOLE (K)

Figure Two

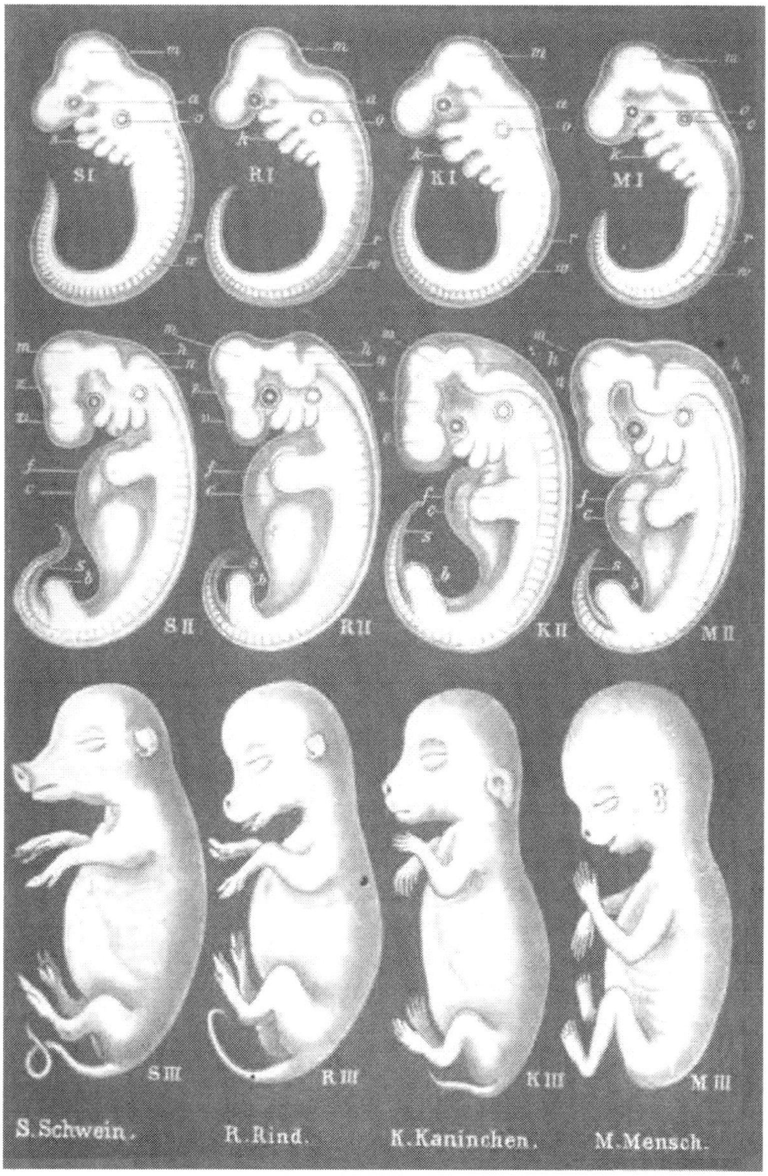

Plate Five from Haeckel's 1874 Edition of *Anthropogenie*
(between pages 256 and 257)

COMPARING PIG, COW, RABBIT, AND HUMAN
EMBRYONIC AND FETAL STAGES OF DEVELOPMENT

Figure Three

> Now, the more complex organisms become, the more evident becomes their inherent kinship. It manifests itself in the absolute and universal uniformity of the basic cellular pattern, and... finally it shines clearly in the general laws of development ('ontogenesis' and 'phylogenesis') which give to the living world, considered as a whole, the coherence of a single upthrust.
>
> *Ibid., pp. 99-100*

> In short, the further the living being emerges from the anonymous masses by the radiation of his own consciousness, the greater becomes the part of his activity which can be stored up and transmitted by means of education and imitation. From this point of view man only represents an extreme case of transformation. Transplanted by man into the thinking layer of the earth, heredity, without ceasing to be germinal (or chromosomatic) in the individual, finds itself, by its very life-centre, settled in a reflecting organism, collective and permanent, in which phylogenesis merges with ontogenesis.
>
> *Ibid., pp. 225-226*

•

What might Haeckel have thought about de Chardin's *The Phenomenon of Man* or the present author's *Intelligent Evolution?* In all probability, Haeckel would have looked at de Chardin's teleology and Omega Point, as well as the present author's paradigm of *intelligent evolution,* as evidence of human vanity and arrogance:

> Since the awakening of the human consciousness, human vanity and arrogance have delighted in regarding Man as the real main-purpose and end of all earthly life, and as the centre of terrestrial Nature, for which use and service all the activities of the rest of creation were from the first defined or predestined by a "wise providence."
>
> *The Evolution of Man, Vol. 2, p. 17*

To Haeckel, there could be no blending, melding, or merging of creationism with evolutionary thinking; he thought an explanation for the totality of life was only an *either-or* prospect:

> Between these two assumptions there is no third course. Either a blind belief in creation, or a scientific theory of evolution.
>
> *Ibid., pp. 36-37*

Additionally, the present author believes that both Lamarck and Haeckel would have chided de Chardin for not including languaging ability as *the major sign* of elevated consciousness through reflection, whose threshold was crossed, according to de Chardin, when primitive man became modern man.

Haeckel asserted that the most important trait that signaled the leap from animal to man is language. When language is coupled with de Chardin's threshold of reflection, an important feature — perhaps the most important feature — of the elevation of consciousness is identified. Haeckel stated:

> The history of the development of languages also teaches us (its Ontogeny in every child, as well as its Phylogeny in every race), that the actual rational language of men developed gradually, only after the body had developed into the specific human form.
>
> *The Evolution of Man, Vol. 2, p. 182*

> Man originated from the preceding stage [i.e., Haeckel's *Ape-men,* or *Alali,* stage] in consequence of the gradual improvement of inarticulate animal sounds into true human articulate speech. Only very uncertain conjectures can be formed as to the time and place of this true "Creation of Man." [brackets mine]
>
> *Ibid., pp. 182-183*

>>>>><<<<<

In closing Section 2.5.4.4, it is important for the present author to add that, like de Chardin, both Lamarck and Haeckel erred in not attributing at least a portion of their evolution-related Weltanschauungen (i.e., world-views) to metaphysics. Had they done so, they would have enabled Christian thinkers to more easily grasp the role of Jesus Christ's consciousness in shaping evolutionary change. Also, many of the ideas, concepts, and constructs of Lamarck and Haeckel that are contested by modern-day evolutionists would be above reproach if they were taken, understood, and applied metaphysically. Unfortunately, Lamarck's theory of use and disuse as well as Haeckel's biogenetic law have been pigeonholed in unflattering ways by people with their own agendas. These people badger others into adopting a narrow way of looking at life. To be sure, their so-called scientifically-correct agendas are often only politically-correct and, for that reason alone, limited in scope.

2.5.4.5 Christianity and Evolution according to de Chardin

Christianity and Evolution is a collection of previously published and previously unpublished essays by Pierre Teilhard de Chardin. It is this collection of essays from which the present author has drawn to write this current section. As a side note for readers who may look into the original essays for themselves, de Chardin is sometimes referred to in commentator footnotes of *Christianity and Evolution* as Père Teilhard, meaning "*Father* Teilhard." (Some readers unfamiliar with French might misconclude that *Père Teilhard* is an alternate spelling for *Pierre Teilhard.*)

De Chardin did what Eddy did not. Eddy stepped into an intellectual and academic abyss of Christian idealism without ever clearly advocating our relationship with a personal Godhead through Christ Jesus or without ever clearly attributing all forms of cosmic, biological, and consciousness changes and maturation to

Christ Jesus. In contrast, de Chardin acknowledged that Christ Jesus — as the principal "centre of universal confluence" *(Christianity and Evolution, p. 87)* — is co-extensive with a relative space-time within which the Creator-God's creation is made ready through physical evolution and cerebralization to perceive the Creator-God as well as to accept and receive Him:

> The fact that Christ emerged into the field of human experience for just one moment, two thousand years ago, cannot prevent him from being the axis and the peak of a universal maturing.
>
> In such a position, finally, Christ, wholly 'supernatural' though his domain may ultimately be, gradually radiates his influence throughout the whole mass of nature. Since, in concrete fact, only one single process of synthesis is going on from top to bottom of the whole universe, no element, and no movement, can exist at any level of the world outside the 'informing' action of the principal centre of things. Already co-extensive with space and co-extensive with duration, Christ is also automatically, [by] virtue of his position at the central point of the world, co-extensive with the scale of values which are spaced out between the peaks of spirit and the depths of matter. [brackets mine]
>
> *Christianity and Evolution, p. 88*

As the present author sees it, authentic Christians have been frozen in time (either by themselves or by the Creator-God) to perceive the Incarnation of Christ as well as to accept and receive the Redemption of Christ. As de Chardin saw it, "our Christology is still expressed in exactly the same terms as those which, three centuries ago, could satisfy men whose outlook on the cosmos it is now physically impossible for us to accept" *(Christianity and Evolution, p. 77)*. Thus, if we are to confront our modern time as well as our future, a correction must be made.

> If we ask in what exactly this correction in relationship consists, the answer must be in bringing Christology and evolution into line with one another.
>
> *Christianity and Evolution, p. 77*

De Chardin added this profound challenge:

> ...if a Christ is to be completely acceptable as an object of worship, he must be presented as the saviour of the idea and reality of evolution.
>
> *Christianity and Evolution, p. 78*

Eddy's writings and de Chardin's writings have little in common, and there is no reason to believe that either author had knowledge of the other's existence although they were contemporaries for almost three decades. What might de Chardin have thought of Eddy and her Christian idealism? The present author has found what he thinks would be representative of de Chardin's thinking about Eddy's systematic theology:

> ...the modern reaction against anthropomorphism has gone much too far, to the point of making us doubt a divine ultra-personality. If we recognize that the true universal (the centre of the universe) cannot, by nature, but be hyperpersonal, then its historical manifestation in a personal form becomes logically comprehensible again, subject to correcting certain of our representations in detail.
>
> *Christianity and Evolution, p. 136*

In contrast, what might Eddy have thought of de Chardin and his convergence of consciousness in an Omega Point through advanced hominization? The present author believes that Eddy would have found de Chardin's thinking on anthropogenesis and hominization as unseemingly anthropocentric. *For example,* the present author believes that Eddy would have derided the following that de Chardin wrote:

> The nineteenth century and the early years of the twentieth were primarily concerned [with throwing] light on man's past — the result of their inquiries [making] it unmistakably clear that the appearance of thought on earth corresponded biologically to a 'hominization' of life. We are now finding that the concentration of scientific researches, focused ahead on the extensions of the 'phenomenon of man,' is opening up an even more astonishing prospect in that direction: that of a progressive 'humanization' of mankind. [brackets mine]
>
> *Christianity and Evolution, p. 140*

As Eddy's universe would be eventually peopled only by transcendent forms in divine Mind, de Chardin's universe would be eventually peopled only by immanent forms vis-à-vis the advanced hominization of "elemental persons" who have become "super-humanized" *(Christianity and Evolution, p. 152)*. Unfortunately, Eddy feared what she did not know or understand concerning physical evolution. (If she had not feared Darwin's theory of evolution, she would not have been so dismissive of important concepts embedded within it.) In contrast, de Chardin embraced what he did not know or understand concerning physical evolution, and he used his fertile imagination, accurate or inaccurate, to provide missing information and constructs to build his paradigm.

For the present author, the human soul — created in the complete image and perfect likeness of the Creator-God — is the wellspring of one's individual creativity. And it is this creativity that is responsible for our imagination and contributes to our individual personality. To be sure, one's personality is an expression of one's organic brain structure, intelligence, experience, education, training, nurture, acculturation, and creativity — all of which inform and construct or deform and destruct one's imagination.

(Many human beings suppress their own imaginations in order to fit in with the culture of their peers. And the imaginations of many others have been warped by those who have imposed constraints on thinking.)

To the present author, the most important paradigm shift provided by de Chardin's thinking is not only in the presentation of the pre-incarnate Christ as the Author of the Genesis Creation but also in the presentation of the pre-incarnate Christ as the Evolver of all life during the Genesis Creation. For de Chardin, Christ's role as Evolver even continued, and still continues, throughout his post-incarnate state:

> Surely this 'Omega Point' (as I call it) is the ideal place from which to make the Christ we worship radiate — a Christ whose supernatural domination, we know, is matched by a physical power which rules the natural spheres of the world. 'In quo omnia constant.' ["In him all things hold together." *Colossians 1:17*] We have here an extraordinary confluence, indeed, of what is given to us by faith and what is arrived at by reason. What used to appear to be a threat becomes a magnificent reinforcement. Far from conflicting with Christian dogma, the boundless dimensional augmentation man has just assumed in nature would thus have as its result (if carried to its ultimate conclusion) a new access of immediacy and vitality to contribute to traditional Christology. [brackets mine]
>
> *Christianity and Evolution, p. 143*

According to de Chardin, evolution was, and is, a part of creation. The following three statements from de Chardin's *Christianity and Evolution* provide helpful working assumptions for Part Three of *Intelligent Evolution:*

1. "Evolution and Christianity coincide fundamentally." *(Ibid., p. 155)*

2. "Evolutionism and Christianity need one another to support and complete each other." *(Ibid.)*

3. "When evolutionism and Christianity are considered in their complementary values, all they call for is the fertilizing and synthesizing of one another." *(Ibid., p. 156)*

Although an understanding of *intelligent evolution* is not required to receive salvation through Christ Jesus, such an understanding is necessary to make sense of all cosmic, biological, and consciousness-related events in the physically-knowable universe: These events all play important roles in setting the stage for: (1) the salvation of souls fallen from the spiritually- or metaphysically-observable universe; and (2) the creation of "a new heaven and a new earth" *(Revelation 21:1 KJV)* at the end of *the Millennium* (the 1,000 year period of time during which Christ Jesus reigns on Earth).

According to the present author's paradigm of *intelligent evolution,* physical creation is the act and physical evolution is the process or method by which that act is achieved. In other words, the physical creation of all life is solely an act of the one and only Supreme Being of *the Whole Universe* (i.e., the God of the Holy Bible), and physical evolution is the major process or method by which the Creator-God has achieved that act. Thus, without fear of insulting, ridiculing, or blaspheming him, we may call Christ Jesus not only Creator but also Evolver, Evolver-Creator, and Creator-Evolver. Contrary to insulting him, acknowledging Christ Jesus as both Creator-God and Evolver-God exalts him because it recognizes who he is and helps to declare his infinite capabilities.

Let us now explore the theoretical physics and cosmology of Stephen Hawking as it relates to the concept of *intelligent evolution.*

2.5.5 Insights, Implications, and Applications from Hawking

Whenever the present author reads articles, monographs, and books on philosophy, theology, or science, he looks to see if he can draw practical metaphysical inferences from their various specialized language labels, statements, principles, and speculations.

In this section, as he did with the writings of Aristotle, Kant, Eddy, and de Chardin, the present author will be integrating his own Christian metaphysical thinking with the thinking of Hawking, primarily by using Hawking's book entitled *A Brief History of Time,* for the purpose of elaborating a Christian metaphysical model of *intelligent evolution* that incorporates modern scientific thinking on cosmology.

Let us now turn to the contributions of Stephen Hawking to clarify some of our own thinking relative to the concept of *intelligent evolution,* especially with regard to cosmic evolution. To be sure, Hawking's literary efforts require and inspire deep thinking through the metaphysical insights they provide as well as the various implications and applications they stimulate — whether or not Hawking intended his efforts to engender metaphysical thinking about theoretical physics, cosmology, and evolution.

2.5.5.1 Hawking's Position on God

In *A Brief History of Time,* Hawking used the noun form of *metaphysics* twice *(pages 9 and 11)* in reference to a philosophical understanding of the initial state of the physically-knowable universe (using the present author's nomenclature) in contrast to a theological, religious, spiritual, or teleological understanding of the universe's beginnings. (On the two pages cited, Hawking specifically contrasted metaphysics with "theology" and "religion.") However, in his chapter entitled *The Origin and Fate of the Universe,* Hawking did introduce teleology when he discussed the *anthropic principle (pages 128 through 131)* as the possible ultimate cause for the physically-observable universe.

Because Hawking's discussion of the *weak* and *strong* variations of the *anthropic principle* rambles and is disjointed, the present author now shares his own perspective on these two variations of the principle:

> (1) The *weak anthropic principle* posits that the physically-observable universe is the way it is so that carbon-based biological life *could/might* evolve.
>
> (2) The *strong anthropic principle* posits that the physically-observable universe is the way it is so that carbon-based biological life *would/must* evolve.

Regardless of variation, the *anthropic principle* acknowledges that the following cosmic prerequisites were necessary for the evolution of carbon-based biological life on Earth:

> (1) four dimensions of space-time (three spatial dimensions and one time dimension);
>
> (2) an expanding universe stable enough to avoid recollapse; and
>
> (3) stable elliptical orbits of planets around a stable solar source (i.e., with pressure balancing gravity) that emits at least some usable energy.

The present author thinks that Hawking would have been loath to admit his pantheism although he might have had an easy time accepting his Aristotelianism. As a theoretical physicist and atheist, how could Hawking's belief system fit the definition of *pantheism?* Hawking sought to replace a need for God with scientific explanations of the observable universe. Instead of looking for Deity in absolute space-time (i.e., in the spiritually-observable universe), Hawking looked for Deity in mathematical and theoretical explanations of matter and relative space-time. For this reason, the present author has coined these three phrases to label Hawking's personal belief system: *mathematical pantheism*, *theoretical pantheism*, and *metaphysical pantheism (metaphysical* used here in its philosophical, and not its spiritual, import). However, just as the Mind of God cannot reside in matter and relative space-time (i.e., as in the classical, immanent sense of pantheism), neither can the Mind of God reside in mathematical derivatives and theoretical elaborations (i.e., as in looking for the Mind of God in empirical data and calculations). Placing one's faith in mathematical solutions and theoretical abstractions is just as pantheistic as placing one's faith in matter. (The present author's previous statement is neither a blanket rejection of *materia medica* nor a denigration of academic study in philosophy or the mathematical sciences.)

Because Hawking acknowledged that the physically-observable universe had a beginning, he also acknowledged that its beginning must have had a "First Cause" but that its "First Cause" could not have been Deity. To be sure, Hawking did not expect to find God at all although he did expect to replace the need for God with scientific explanations of heretofore unexplained natural phenomena. Because Deity as "First Cause" is a reasonable explanation for anything that has had a beginning, Hawking would have preferred that the physical universe did not have a beginning at all.

In fact, Hawking sought proof that, although finite, the universe has no boundary and, therefore, did not have a beginning. Why? It made Hawking personally uncomfortable to include Deity as a possible explanation for the physically-observable universe's beginning:

> It would be very difficult to explain why the universe should have begun in just this way [in a hot big bang], except as the act of a God who intended to create beings like us. [This is a restatement of the *anthropic principle.*] [brackets mine]
>
> *Ibid., Kindle Locations 1784-178*

Hawking was so uncomfortable with the possible existence of Deity that he put forth a proposal "for aesthetic or metaphysical reasons" *(Ibid, page 142)* that space-time did not originate from a hot big bang and, therefore, had no beginning (and, consequently, no boundary):

> So long as the universe had a beginning, we could suppose it had a creator. But if the universe is really completely self-contained, having no boundary or edge, it would have neither beginning nor end: it would simply be. What place [is necessary], then, for a creator? [brackets mine]
>
> *Ibid., Kindle Locations 2011-2013*

Casually reading *A Brief History of Time,* one might conclude that Hawking believed that there is a God. *For example,* he stated:

> God, being omnipotent, could have started the universe off any way he wanted. That may be so, but in that case he also could have made it develop in a completely arbitrary way. Yet it appears that he chose to make it evolve in a very regular way according to certain laws.
>
> *Ibid., Kindle Locations 231-233*

> These laws may have originally been decreed by God, but it appears that he has since left the universe to evolve according to them and does not now intervene in it.
>
> *Ibid., Kindle Locations 1709-1710*

> The whole history of science has been the gradual realization that events do not happen in an arbitrary manner, but that they reflect a certain underlying order, which may or may not be divinely inspired.
>
> *Ibid., Kindle Locations 1714-1716*

> ...if we do discover a complete theory [i.e., a unified theory of everything], it should in time be understandable in broad principle by everyone, not just a few scientists. Then we shall all, philosophers, scientists, and just ordinary people, be able to take part in the discussion of the question of why it is that we and the universe exist. If we find the answer to that, it would be the ultimate triumph of human reason — for then we would know the mind of God. [brackets mine]
>
> *Ibid., Kindle Locations 2603-2606*

To Hawking, it would seem that the concept of Deity is only a space-filler (pun intended) for areas (1) that science does not yet understand and/or (2) that have not yet been expressed by mathematical equations or empirical data from observations or experimental investigations.

Hawking tried to shed light concerning his position on God in a chapter entitled *Is There a God?* in his book *Brief Answers to the Big Questions:*

> I prefer to think that everything can be explained... by the laws of nature.
>
> *Brief Answers, page 26*

> I use the word "God" in an impersonal sense, like Einstein did, for the laws of nature, so knowing the mind of God is knowing the laws of nature.
>
> *Ibid., page 28*

> I have no desire to offend anyone of faith, but I think science has a more compelling explanation than a divine creator.
>
> *Ibid., page 34*

> The role played by time at the beginning of the universe is, I believe, the final key to removing the need for a grand designer and revealing how the universe created itself.
>
> *Ibid., page 37*

When Hawking stated that "there is no possibility of a creator because there is no time [before *the hot big bang]* for a creator to have existed in" *(Ibid., page 38, brackets mine),* he demonstrated his ignorance that the Creator-God resides in eternity (i.e., in absolute space-time) and not in infinity (i.e., in relative space-time). Based on the mathematical perspective that all of the positive energy and negative energy in the physically-knowable universe add up to zero, Hawking deduced that "if the universe adds up to nothing, then you don't need a God to create it" *(Ibid., page 33).* His observation helps to prove the present author's point that Hawking's personal belief system had its underpinnings in *theoretical pantheism* because Hawking concluded that, if God were to exist, His existence: (1) could only be in the physically-knowable universe and (2) could only be proved through mathematics and empirical data.

Of course, we must not be dismissive of Hawking's true genius, but we must also qualify that his genius was in the realm of the physically-knowable and not in the realm of the spiritually-knowable. Perhaps Hawking was influenced by Satan's double whammy: (1) uncanny intellectual ability coupled with (2) vanity.

The Bible provides this Christian response to Hawking:

> Know that the LORD, He is God; It is He who has made us and not we ourselves. [The universe did not create itself.]
>
> *Psalm 100:3a KJV [brackets mine]*

We are comparing physical apples to metaphysical galaxies when we try to find the Mind of the Creator-God in the physically-knowable universe. *For example,* where does human happiness come from? Human happiness comes from hope. Hope comes from faith. And faith is imparted to us through hearing the gospel message of salvation. When we hear the gospel message of salvation, faith is implanted within our souls and watered by spiritual truth from the Creator-God's Holy Spirit. Spiritual happiness, joy, peace, contentment, faith, and fulfillment are nowhere to be found in the physically-knowable universe. However, they are found everywhere in the spiritually-observable universe (except in the Lake of Fire).

The present author believes that Hawking was afraid of God. Perhaps Hawking's formidable physical disabilities made him afraid of God or made him conclude that no God would have permitted him to suffer the indignities associated with his ALS (commonly known as Lou Gehrig's Disease).

In the final analysis, the present author believes that Hawking vacillated between believing in God and not believing in Him. If Hawking really did not believe in God, he would not have devoted *any* time at all to discussing the *anthropic principle.* In his chapter entitled *The Origin and Fate of the Universe,* Hawking discussed the *anthropic principle* as the possible purpose for the physically-observable universe. To the present author, Hawking's discussion proves that he was greatly conflicted by the concept of a personal Deity. No atheistic scientist (as Hawking labeled himself) would ever discuss the *ultimate purpose* for natural events. Atheistic scientists seek to explain *how* but not *why* something exists. Atheistic scientists might agree that there should be purpose and meaning *in* each person's life through self-survival and individual involvement in meaningful activities, but atheistic scientists would not look to identify the individual or collective purpose and meaning *for* life. In order to maintain credibility and objectivity, even most Christian biologists would be careful not to ascribe purpose for specific organismic evolutionary developments when addressing the members of an academic scientific society.

2.5.5.2 Understanding Cosmic Theory

Although Aristotle concluded that the Earth is round based on his observations of the Earth's shadow cast on the moon during a lunar eclipse, he did not know if the Sun or the Earth was the center of our solar system. Further, he did not know if the material universe was static (i.e., its heavenly bodies fixed and stationary) or if its stars moved away from or toward one another. "But in 1929, Edwin Hubble made the landmark observation that wherever you look, distant galaxies are moving rapidly away from us [and most are moving away from each other as well]. In other words, the universe is expanding" *(A Brief History of Time, page 9, brackets mine).* If Eddy had been alive in 1929 (she died in 1910), the present author is sure that she would have picked this scientific fact to illustrate metaphysically that human consciousness is always expanding to fulfill the expectations of a divine Mind. And, although de Chardin was alive in 1929 (he died in 1955), the present author is not aware of his thoughts on the matter but can imagine that the fact of an ever-expanding universe might well figure into de Chardin's conceptualization of the Omega Point — toward which, de Chardin concluded, the universe and human consciousness were always evolving.

It should be noted here that, although the physically-observable universe continued — and still continues — to expand since the time of its beginning, there are irregularities, singularities, and exceptions to its expansion: *for example,* in the contraction of the central regions in some disklike rotating galaxies and in the collapse of some stars to form black holes. Metaphysically speaking, these exceptions should help to remind us that causes and explanations of causes for human events also do not fit the "one size fits all" model. We need to remind ourselves that causes and explanations for human events are multivariate in nature (i.e., in essence) with some irregularities, singularities, and exceptions.

At the end of his life, Hawking looked for the elusive theory of everything (TOT) that would completely unify quantum mechanics with Einstein's general theory of relativity concerning gravitation — which two partial theories do not easily fit with one another "to provide a single theory that describes the [entire physically-observable universe]" *(Ibid., page 11, brackets mine).* Hawking believed that it is the ultimate goal of science to produce *a quantum theory of gravity* that would provide a model of the universe capable of predicting the behaviors of both (1) macroscopic bodies that are involved in gravitational interactions and (2) atoms, subatomic particles, and radiation that appear not to be involved in gravitational interactions.[23] To the present author, a unifying and complete theory would just simply acknowledge that matter behaves differently based on its mass or lack thereof (i.e., as in the masslessness of a photon). To the present author, a unifying and complete model would just simply hold (i.e., conceptualize) the entire physically-observable universe while simultaneously attending to the two primary partial theories that describe the behavior of its matter in different conditions — *for example:* (1) agglomerated, mixed, and extraordinarily vast and (2) non-agglomerated, unmixed, and extraordinarily tiny.

Every undergraduate student in inorganic chemistry is taught the difference between true solutions and true suspensions. A true solution is composed of a dispersing medium containing dispersed particles whose diameter is less than 1 nanometer (.001 micrometer or 1 millimicrometer), and a true suspension is composed of a dispersing medium containing particles whose diameter is greater than 1,000 nanometers (1 micrometer or 1,000 millimicrometers). Dispersed particles in a true solution remain dispersed indefinitely

[23] The so-called *quantum theory of gravity* greatly depends on the collapsibility of all forces in quantum mechanics.

because they remain dissolved: the dissolved particles are too small for gravity to act upon. But dispersed particles in a true suspension are dispersed only momentarily and settle out (i.e., form a sediment) because they are never truly dissolved in the dispersing medium: the suspended particles are large enough for gravity to act upon. The unifying and complete explanation for these two behaviors is in understanding that matter behaves differently based on its particle size (i.e., diameter). Similarly, for the present author, the unifying theory for quantum mechanics and Einstein's general theory of relativity is found in understanding that matter behaves differently based on (1) the size of its mass or (2) masslessness. In this way, although the two partial theories are not completely unified by identifying one common force, they can be understood stereoscopically.

Of course, employing Christian metaphysics, it would be understood that the supernatural force and power of the Creator-God provides the ultimate — unified and complete — model for understanding cosmic evolution, biological evolution, and consciousness evolution in the physically-observable universe. Hawking would conclude that the present author has turned to this supernatural explanation because he (the present author) does not know the natural law that governs both gravitation and quantum mechanics. In response, the present author would advise theoretical physicists and cosmologists to continue to look for a complete theory that unifies gravitation and quantum mechanics in the physically-observable universe at the same time that they give assent to the creative force and power of divine Mind in *the Whole Universe*. In other words (because reiteration is the mother of learning), the Creator-Evolver Himself is the complete unified Principle that blends, melds, and harmonizes the general theory of relativity with the wave/particle duality of quantum mechanics[24] because He alone provides the supreme unification energy that holds *the Whole Universe* together.

---

[24] The wave/particle duality of quantum mechanics acknowledges that waves sometimes act like particles and that particles sometimes act like waves.

The intellectual unification in a complete theory of general relativity's gravitation with quantum mechanics is as difficult as the intellectual unification in a complete theory of one's self-worth through self-value with one's self-effacement through personal humility. Standing up for yourself by never permitting yourself to be treated as less than you are can only be reconciled with personal humility through the force-carrying power of Christ Jesus in self-sacrifice.

The metaphysical value in understanding the concept of unifying partial theories is found in achieving reconciliation of various Bible truths that seem to be irreconcilable with one another — *for example,* reconciling: (1a) the truth that we are saved by faith and not by works with (1b) the truth that faith without works is a dead faith; (2a) the truth of one's spiritual entitlement with (2b) the truth of the spiritual requirement for obsessive gratitude in oneself; (3a) declarations of spiritual truth relative to the betterment of one's human conditions with (3b) praying for the Creator-God to change those conditions only if it is His Will that they be changed; and (4a) the benefits of healthy sexual desire with (4b) the benefits of celibacy. (Each pair of the four sets of "partial truths" presented here can be completely unified; thus, no true dilemmas exist for any of them.)

Where cosmology, theoretical physics, and Christian metaphysics all come together is in their understanding of relative time. Cosmology, theoretical physics, and the present author's brand of Christian metaphysics agree that there was no relative time before the beginning of the physically-knowable universe. According to Hawking: (1) "Time did not exist before the beginning of the universe;" (2) "The concept of [relative] time has no meaning before the beginning of the universe;" and (3) "One may say that [relative] time had [its] beginning at the big bang" *(Ibid., pages 8 and 9, brackets mine).*

Although Aristotle and Isaac Newton believed in the existence of absolute time in the material universe, there can be no absolute time in a universe of relative space-time (i.e., the physically-observable universe): Einstein's "theory of relativity put an end to the idea of absolute time" *(Ibid., page 22).* Absolute time only exists in the spiritually-observable universe, where time *is* space and space *is* time. *Time* is defined differently in the spiritually-observable universe and in the physically-observable universe. Moreover, in the physically-observable universe, light travels at a finite speed (velocity) and is measured in units of distance divided by units of relative time. Concerning light in the physically-observable universe, Hawking reminded us that (1) "only light or other waves that have no intrinsic mass can move at the speed of light," and that (2) "nothing can move faster than the speed of light" *(Ibid., page 21).* In the spiritually-observable universe, the speed of the Creator-God's light (i.e., His Glory) is at an eternal speed in absolute time because that universe is dimensionless. In other words, the speed of the Creator-God's light is immeasurable because there are neither units of relative distance nor units of relative time in the spiritually-observable universe. Distance, speed, and relative time do not exist in the spiritually-observable universe. (Absolute time does exist as *folds* in the absolute space of the spiritually-observable universe.)

An important metaphysical comparison is that, just as nothing physical can travel faster than the speed of light in the physically-observable universe, so is it also true that the light of the Creator-God travels faster than the speed of light everywhere in *the Whole Universe* — which includes the physically-observable universe and the spiritually-observable universe together. (The present author acknowledges that the phrase *faster than* is a relative descriptor that has meaning only for those in the physically-observable universe.) Indeed, when the Creator-God chooses to appear in the physically-observable universe, there is no lag in time between His decision to appear and His actual appearance.

An *event* for Hawking is defined as "something that happens at a particular point in space and at a particular time" *(Ibid., page 24).* This definition is in keeping with what was told to the present author from Heaven in 1966 that "time [in the physically-observable universe] is a sequence of related events." It is also in keeping with how the Holy Bible measures events on Earth — which is to say, *relative to one another*. To be sure, any and all Biblical prophetic events that will be fulfilled in the Earth's future take place in the relative space-time of the physically-observable universe and not in the absolute space-time of the spiritually-observable universe, the latter of which is the place (or the state) where there are only experiences and not events (based on Hawking's definition of *event,* of course). In the physically-observable universe, time is not absolute; time is relative. Therefore, time associated with the events of the Biblical creation is relative to the observer. Thus, to one observer, Biblical creation might be seven days; to another observer, Biblical creation might be seven billion years.

One of Hawking's most important statements in *A Brief History of Time* is: "we must accept that time [in the physically-observable universe] is not completely separate from and independent of space but is combined with it to form [a four-dimensional] object called space-time" *(Ibid., page 23, brackets mine).* This provides a metaphysical analogy to the combined absolute space and absolute time that exist in the spiritually-observable universe, where there are no events but only experiences in consciousness. (The folds of absolute time in the absolute space of the spiritually-observable universe are analogous metaphysically to wormholes[25] in the physically-observable universe.)

Only events in the future of the Biblical creation can be affected by what happened at the Biblical creation because nothing can travel faster than light in the physically-observable universe — except, of

[25] **_Wormholes_ are thin tubes of space-time in the physically-observable universe that (1) connect distant regions of space that are far apart from one another and (2) permit time travel (i.e., traveling faster than the speed of light) between those regions.**

course, for (1) the Creator-God's light and (2) time travel through wormholes. The events that do not lie directly in the future or directly in the past of the Biblical creation are said to lie in the *elsewhere* of the Biblical creation. Thus, what has happened and what will happen in Heaven relative to the Biblical creation of the physically-observable universe lies in the *elsewhere* of its creation.

The present author believes that the paradigm of *intelligent evolution* provides the perfect explanation for understanding the *elsewhere* of Biblical creation in the existence of spiritually-observable and physically-observable parallel universes. The paradigm of *intelligent evolution* provides (1) a perfect law, or theory, that fits all of the observations and (2) a just model that predicts our ultimate future with certainty.

When people on Earth look at the star-filled sky today, they are viewing a past that only has relevance to them by providing them with light from millions and billions of years ago (depending on the distance that the stars are now from Earth). *For example,* some of the starry light we see today has traversed as much as forty-eight billion light-years to reach us. (In a sense, we are witnessing an afterglow from *the Big Bang*.) To be sure, the stars that emitted light forty-eight billion years ago may no longer exist today; and light from them that reached the Earth on the fourth "day" of Biblical creation *(Genesis 1:14-19)* — when the Bible records that light from the stars first arrived in our solar system — may have traveled as far as forty billion light-years, Earth having been formed approximately four to five billion years ago and the physically-observable universe not having expanded as far as it has today.

Note: Although Earth is not at the center of the entire physically-observable universe and is actually off-center (but not at its edge), Earth is at the center of Earth's observable universe. In other words, the vastness of the physically-observable universe only leads us to believe we are near its center. To be sure, there is an exact center of the physically-observable universe, but we are not in it. (Although an exact center exists, its coordinates are no longer plottable.) For the sake of further clarification: because the ever-expanding physically-observable universe originated from *the Big Bang,* and because Earth is off-center in the physically-observable universe, some starry light has traversed as far as forty-eight billion light-years to reach us today (one-half of the full ninety-six billion light-years that represent the diameter of the entire physically-observable universe today).

Perhaps the stars of the physically-observable universe have already disappeared and their disappearance will not be seen by the naked eye until the following prophesied events occur toward the end of *the Millennium,* recorded in the Bible as the time when "heaven departs like a scroll that is rolled together..." and "the stars of heaven fall to Earth — even as a fig tree casts her unripened fruit when shaken by a mighty wind" *(Revelation 6:14 and 6:13 KJV Paraphrase).* (To be sure, the indicated "falling stars" could, instead, be debris from comets, asteroids, and/or meteoroids.)

These end-time events were also spoken of by Christ Jesus:

> Immediately after the tribulation of those days shall the sun be darkened, and the moon shall not give her light, and the stars shall fall from heaven, and the powers of the heavens shall be shaken.
>
> *Matthew 24:29 KJV*

> And the stars of heaven shall fall, and the powers that are in heaven shall be shaken.
>
> *Mark 13:25 KJV*

These prophesied changes are consistent with Einstein's general theory of relativity, which "implied that the universe must have a beginning and, possibly, an end" *(A Brief History of Time, page 35).* However, let us be reminded that, although the physically-observable universe is changing and will continue to change, the spiritually-observable universe is unchanging and will remain unchanged. Although we were changed when we fell from Eden, Eden remained unchanged. (Eden was not even changed by our absence from it.)

In the physically-observable universe, time and light are influenced by gravity. (1) Gravity slows time down; and (2) much like water refracts visible light and any objects in water are not exactly where they appear to be *(for example,* in spear fishing),[26] the gravity of massive objects in space bends visible light. For the latter reason, some stars are not exactly where they appear to be from Earth (i.e., some stars are more dislocated than others because of the gravitational interactions through which their light has been bent). In contrast, time and light in the spiritually-observable universe are absolute and eternal, they do not change, and they are not influenced by gravity. Why? (1) There is no gravity in Heaven; (2) time does not pass in the ecstasis of Heaven; and, (3) although massless, divine light has no particle/wave duality. (Perform a word search for *theions* in this book for additional insights on divine light, or the Creator-God's Glory.)

In the physically-observable universe, stars have their own luminosity because of nuclear fission and fusion reactions deep within them and at their surface. Metaphysically speaking, *stars* in the spiritually-observable universe are either unfallen created beings or restored fallen created beings whose luminosity is from the Creator-God's divine light, which they transmit and reflect. (In addition, saved human beings appear luminous to incorporeal beings in the spiritually-observable universe, depending on how close they are to the Creator-Savior in their own daily faith walks.) In the spiritually-observable universe, the stars of the Creator-God all belong to a metaphysical syncytium in His Supraconsciousness.

[26] **The refraction of light at the interface between air and water is an electromagnetic phenomenon and not a gravitational phenomenon.**

> Who laid the cornerstone of [the material universe], when the morning stars sang together, and all the sons [heirs] of God shouted for joy? [brackets mine]
>
> *Job 38:6b-7 KJV Paraphrase*

> And they that are wise shall shine as the brightness of the starry sky; and they that turn many to righteousness shall shine as the stars forever and ever.
>
> *Daniel 12:3 KJV Paraphrase*

Although there is no boundary at the outer edge or horizon (i.e., fringes) of the physically-observable universe, one space-time boundary, so to speak, is found at *the Big Bang* when relative space-time began. From the time of *the Big Bang* onward, the physically-observable universe has continued to expand. How do we know that it still continues to expand? "If the source [of visible light from the stars] is moving away from us, the wavelength of the [light] waves we receive [from the stars] will be longer [that is, on the longer wavelength side of the visible light spectrum]. In the case of [visible] light, therefore, this means that stars moving away from us will have their spectra shifted toward the red [and dark red] end of the spectrum (red-shifted)" *(A Brief History of Time, page 41, brackets mine).* Hubble found that, because most galaxies are red-shifted, (1) they are moving away from us and (2) those furthest from us are moving away the fastest. Commenting on Hubble's discovery, Hawking stated "that the universe is expanding was one of the great intellectual revolutions of the twentieth century" *(Ibid.).* Indeed, this discovery has many implications. One implication is that the entire physically-observable universe will probably not end in a *Big Crunch* (i.e., in a massive contraction like a rubber band snap). Even the physically-observable universe's own gravitational attractions will never stop its expansion: Neither the combined masses of its stars nor the sum of its dark matter is sufficient to halt the universe's expansion. To be sure, this is not inconsistent with the Biblical prophecy that, at the end of relative space-time, all "elements shall melt with fervent heat" *(2 Peter 3:10*

& 12 KJV). The Bible further teaches us that, one day, the Creator-God Himself will infill the physically-observable universe with the totality of His power and energy *(1 Corinthians 15:28).*

Note: Although there is no outer physical barrier to the physically-observable universe, the velocity of its outward expansion (albeit accelerating) is a limiting factor and, thus, imposes a space-time boundary of sorts.

In the Bible, the Creator-God mentioned what happened before the beginning of creation when He referred to Lucifer's Fall *(Isaiah 14:12-14).* Hawking also understated what happened before the beginning of the universe:

> As far as we are concerned, events before the big bang can have no consequences, so they should not form part of a scientific model of the [physically-observable] universe. We should therefore cut them out of the model and say that [relative] time had [its] beginning at the big bang. [brackets mine]
>
> *A Brief History of Time, page 49*

Regardless if the physically-observable universe started with a bang or just simply appeared, it had a beginning. The present author believes that its beginning happened synchronously with Lucifer's Fall, which introduced temporality into a minuscule speck of *the Whole Universe* (see Figure One, entitled *The Outpocketing of Temporality from Eternity,* on Page I-23).

In order to think from a Christian metaphysics standpoint, we need to exchange the quantum theory of gravity with the quantum theory of gratitude. In physics, strength of gravity determines how close one heavenly body is to another. In Christian metaphysics, depth of gratitude determines how close one is to the Creator-God. For saved human beings, gratitude always begins and ends with gratitude for creation, salvation, and the indwelling presence of the

Creator-God's Holy Spirit. So, if you want to grow closer to the Creator-God, express gratitude to Him for all that He is and for all that you have (even if it is only a wheelchair).

>>>>><<<<<

In concluding their reading Part Two of *Intelligent Evolution*, students of Christian metaphysics should assume that the words *creation* and *evolution* are synonymous when they are referring to the same Christ-empowered and Christ-driven *becoming* in the origin and mutability of living things. Indeed, the paradigm of *intelligent evolution* provides the one unifying law that demonstrates the simplicity, complexity, and power of Christ Jesus as Creator-Evolver in the purpose and diversification of his entire physical creation.

Part Three

The Theory of Intelligent Evolution: Explaining the Solution to the Problem

3.1 Elaborating a Cohesive and Coherent Metaphysical Theory of Intelligent Evolution Incorporating the Creator-Savior

The present author has transitioned to using *Creator-Savior* rather than *Creator-God* for Part Three of *Intelligent Evolution* because its readers and listeners have been sufficiently prepared by Parts One and Two to understand that the God of the Holy Bible is both Creator and Savior. To be sure, *God the Father* and *God the Son* intersect in eternity at the center of creation as well as at the full circumference of salvation (this is a metaphysical description of their unity). There is no Creator other than the God of the Holy Bible. There is no Savior other than the God of the Holy Bible. And there is no Scripture other than the Holy Bible. The appellation *Creator-Savior* best represents Who the God of the Holy Bible really *is*.

What has been learned from Parts One and Two of this book will now be imported to Part Three in order to build a cohesive and coherent metaphysical theory that extols the creativity of the Lord God Almighty in His *intelligent evolution* of biological life on Earth. Considering the potential accusation that the present author is insulting the Lord God Almighty by proposing a paradigm of *intelligent evolution,* his response is that, in fact, it is insulting to the Lord God Almighty for human beings to refuse to consider that *intelligent evolution* is one of the infinite number of processes, methods, and means that the Creator-Savior had available to Him to originate and diversify biological life in the sequenced, staged events recounted, but only summarized, in Chapter One of Genesis. Indeed, the present author believes that if the Creator-Savior were not Who He is, He would be entirely exasperated by human beings forever insulting Him by telling Him what He can do, should do, or did not do as well as what He means, should mean, or did not mean.

The sender, or author, of a message who controls the presentation of a paradigm and its explanations controls the acceptance of the paradigm by its prospective recipients (i.e., readers, listeners, and viewers) — of course, depending on their intellectual capacities and spiritual capabilities. To be sure, the intellectual capacities and spiritual capabilities of prospective recipients of a message are determined by their individual nurturance, training, education, self-discipline, desires, interests, and needs. Thus, ignorance of a particular paradigm is partly determined by the lack of exposure to a sufficiently-articulated paradigm from a sender as well as partly determined by the lack of an inquiring mind in a prospective recipient. In other words, prospective recipients of an intended message cannot be convinced of something that has never been sufficiently-articulated to them or something in which they have no interest. Therefore, not only does an author have the responsibility to sufficiently-articulate his or her message, the prospective recipient has the responsibility to be sufficiently-prepared, sufficiently-interested, and sufficiently-motivated to receive it.

The paradigm of *intelligent evolution* will only appeal to those who have the necessary intellectual capacities and spiritual capabilities accompanied by the necessary desire to understand it. Moreover, the paradigm of *intelligent evolution* will only appeal to those who would like to blend, meld, harmonize, and synthesize certain theological and philosophical principles regarding creation with certain scientific principles regarding cosmic evolution, biological evolution, and consciousness evolution using metaphysics as a tool. Because the present author has spent his entire life seeking to understand and articulate this paradigm, he does not expect students to quickly understand it. From the present author's perspective, rapid spiritual progress and understanding are always suspect. Indeed, all spiritual progress and understanding — as well as their elaboration for others — take a long time relative to one human lifetime.

3.2 At the Beginning

3.2.1 Space-Time Warping and Gravitation

Absolute space-time exists only in eternity, where space and time are perfectly united in *an eternal now.* Therefore, absolute space-time exists only in the spiritually-observable universe. Absolute space-time does not exist in the physically-observable universe. Instead, relative space-time exists in the physically-observable universe. And, although they are melded in the physically-observable universe as three spatial dimensions plus one temporal dimension, relative space and relative time are distinguishable from one another.

Relative time is a sequence of related events. Relative time is neither a thing nor a thing-in-itself. Relative time is defined by what happens in it. Relative time has no meaning without events. If nothing happens, there is no relative time. *For example,* nothing happens beyond the fringes of the physically-observable universe (i.e., its cosmic horizon); therefore, relative time does not exist there. Of course, neither does absolute time exist there. The empty vacuum of space beyond the cosmic horizon is devoid of all time.

Relative space has no time of its own. Relative space curves to accommodate mass, and mass brings time with it. Because mass brings time with it, all mass ages. Relative space does not age. The only reason that all corporeal beings die is because they have mass. Regardless if corporeal beings are weightless in relative space or not, they still have mass and eventually die. In contrast, incorporeal beings do not die because they are not only weightless but also massless.

Because each corporeal entity has a spirit (spirit is a function, or divine property, of consciousness), all corporeal entities continue to exist after their physical forms die. Because each human being has a soul in addition to a spirit, its consciousness has additional components. Although other corporeal beings each have a spirit, they do not have souls and, therefore, are not able to make moral and ethical decisions nor demonstrate higher order creativity. Because entities with souls were made in the expressed image and perfect likeness of the Creator-Savior, entities with souls are able to do what entities without souls are unable to do.

Without mass, energy, or particulate motion of any kind, the relative space beyond the physically-observable universe is dimensionless. It is dimensionless even though it can accommodate objects that have spatial dimensions, relative time, and movement. Relative space beyond the cosmic horizon is metaphorically analogous to eternity in that eternity also has no physical dimensions.

Relative space is not just *nothing* in a metaphorical or metaphysical sense; relative space is *nothing* in the truest sense of the word. So, technically, the Creator-Savior did not create *something out of nothing* when He created the physically-observable universe; He created *something in nothing.* Also, technically, relative space cannot curve or warp because it is not a true fabric in the physically-observable universe despite mathematical paradigms that might imply that it is; only mass and energy, and their motions and interactions, can curve or warp. However, relative space can accommodate all dimensions of mass and energy, their motions and interactions, and their curving or warping. It is only in this accommodation that relative space curves or warps.

When Christ Jesus multiplied the loaves of bread and fish, he did so as Creator-Savior using metaphysical mathematics. Souls constrained by human bodies cannot employ metaphysical mathematics but the Creator-Savior can in response to their prayer requests to him because he remains in operation in the physically-observable universe through his Holy Spirit. Thus, that prayer requests are required sets limits to the power of personal declarations used to effect change in corporeality. This is not to say that declarations should not be used to effect change; this is to say that declarations should only be used in combination with prayer requests to the Creator-Savior.

Relative time began when the events of *the Big Bang* occurred. A rift in absolute space-time occurred because of *the Luciferian Fall*. This rift resulted in *the Big Bang*. The Creator-Savior simultaneously used the explosion of *the Big Bang* to create something out of nothing: the nucleosynthesis of particulate matter from energy. (Here, the word *nothing* in the phrase *something out of nothing* is not referring to the nothingness of space.) He then created gravity as a by-product of the warping of relative space-time to accommodate mass (the warping occurred as subatomic energy aggregated to produce matter and matter aggregated to produce mass).

Gravitation is not only the universal field that makes our world go round, it is the universal interaction through which the Creator-Savior creates in the physically-observable universe. Gravitation is the interaction that the Creator-Savior used to evolve the physical universe in these four ways: (1) cosmologically (including the formation of galaxies, galaxy clusters, and cosmic abiotic chemicalization); (2) geologically (including the formation of planets, solar systems, and planetary abiotic chemicalization); (3) biologically (including the emergence of biological life from nonliving matter and that life's capability for complexification); and (4) neurologically (including the gradual formation of a highly developed central nervous system and the expanding expression of consciousness in corporeality through that system). To be sure, the evolution of a complex nervous system is the pinnacle of all physical evolution because it permits the expression of moral, ethical, and higher order creative consciousness in the physically-observable universe by fallen beings that possess souls.

All physical evolution is the effect of gravitational interaction (including its elusive abstraction, quantum gravity) across a warped and warping relative space-time. Gravitational interaction is what the Creator-Savior used to create in a five-dimensional universe. (In its simplest terms, the physically-observable universe is a five-dimensional universe with its fifth dimension including subatomic interactions where things can be in two or more places at once).

The Creator-Savior evolved gravitation from the events of *the Big Bang*. Those events warped the newly-formed relative space-time to accommodate accumulating mass and create gravitation. The Creator-Savior warped relative space-time in the production of mass through the aggregation of matter. This is paradoxical because the Creator-Savior then used gravitational interaction for the rest of physical evolution after the initial moment of *the Big Bang.* (In other words, the Creator-Savior created gravitation in the physically-observable universe to further elaborate on His creation of mass.)

The Creator-Savior controls gravitational interaction across relative time. Gravitational interaction does not control the Creator-Savior. The Creator-Savior uses gravitational interaction. Gravitational interaction does not use the Creator-Savior. Gravitation is an interaction across relative time, and physical evolution is a process over relative time. Gravitation and physical evolution do not compete with the Creator-Savior and His ability to create. The Creator-Savior originated gravitation to help mold the physically-observable universe through the process of physical evolution. Physical evolution does not compete with creation. In fact, the Creator-Savior used the process of physical evolution to create in the physically-observable universe. The Creator-Savior did not just *allow* physical evolution to take place; the Creator-Savior *directed* all physical evolutionary processes.

Relative time speeds up and slows down in relation to gravitation. In physics, this is known as *gravitational time dilation.* Relative time moves more slowly the closer one is to a gravitational field and faster the farther one is away from a gravitational field. This helps to explain why (albeit not entirely), when angelic beings willingly step from their absolute space-time in eternity to relative space-time in temporality, their electromagnetic forms (their *astral gelatinous™* somatic identities) are decelerated, or *stepped down,* upon entering a gravitational field, thereby manifesting as human life forms on the planet Earth. (This does not explain the process entirely because such a transformation of angelic beings also remains under their individual conscious control.) During *the Millennium,* souls in Heaven will also be able to manifest visibly on Earth, albeit not in corporeal forms but in their translucent *astral gelatinous™* forms.

If relative time is a sequence of related events (and it is), then a physical causality is implied. Thus, because it is physical, such a causality can be explained by the laws of physics. One of the reasons that some contemporary physicists question the validity of *the Big Bang* is because they understand that, in accepting its singularity, they would need to also accept the existence of a different causality prior to the beginning of the current physical reality (more specifically, they might need to accede to the existence of a Prime Mover that existed prior to *the Big Bang).* In other words, they know intuitively that *the Big Bang* proves the existence of a reality other than a physical reality governed by the laws of physics. However, because their belief systems will not permit them to accept the existence of a spiritual reality, their belief systems require them to look for ways to reject *the Big Bang* explanation for the existence of the physically-observable universe. (For the sake of clarification, the noun phrases *Prime Mover, Supreme Being, Creator-God,* and *Creator-Savior* all refer to the God of the Holy Bible and to no other.)

Many physicists today conceptualize the physically-observable universe as a membrane that resides in a hyperspace (often referred to by them as a "brane" existing in a higher dimensional "bulk"). The end of the physically-observable universe (its destruction and dissolution) and the beginning of "a new heaven and a new earth" *(Revelation 21:1 KJV)* does not occur because of an implosion (the reverse of *the Big Bang* often referred to as *the Big Crunch);* the end occurs because of the physically-observable universe's collision with forces and fields not part of its membrane, specifically when the Creator-Savior's Supreme Being intentionally infuses the physically-observable universe at the end of the millennium of Jesus Christ's reign on Earth. Before that infusion, the Creator-Savior's Supreme Being resides in the hyperspace that the present author generally refers to in his writings as *the spiritually-observable universe.*

There are no random accidents in the Creator-Savior's *intelligent evolution* of the physically-observable universe because everything proceeds according to His divine plan. To be sure, there are no *nonrandom* accidents either because, if they are *nonrandom,* they cannot also be *accidents.* Accidents in the paradigm of neo-Darwinism are replaced in the paradigm of *intelligent evolution* by the Creator-Savior's directed chance that results in physical events that have purpose and meaning. In the paradigm of *intelligent evolution,* the closest that one can come to the notion of accidents is in *coincidental anomalies,* which are inconsequential and neutral in their import to, and bearing on, *intelligent evolution* and its use in the Creator-Savior's Plan of Salvation for the rescue of errant souls and the creation of "a new heaven and a new earth" at the end of *the Millennium.*

3.2.2 Displacement Theory

For the present author, building a metaphysical theory of *intelligent evolution* utilizes *displacement theory* to explain: (1) the origin of matter in relative space-time (i.e., temporality); and (2) the origin of eternal souls in corporeality (i.e., physicality).

3.2.2.1 The Origin of Matter in Relative Space-Time

The first aspect of *displacement theory* is the origin of matter in relative space-time. This aspect includes the supposition that an infinitesimally small amount of spiritual energy (i.e., a minuscule number of *theions)* was first displaced from eternity to temporality at the time of Lucifer's rebellion (also referred to as *the Luciferian Fall).* This displacement of spiritual energy resulted in what is commonly referred to as *the Big Bang.* For the purpose of *intelligent evolution,* a *theion* is "the smallest indivisible unit of divine, eternal, or spiritual energy." (It is this new sense for the already-established Greek word **θεῖον** *theion* [G2303] that causes its anglicized form to be a *neologism.*) An analogy that might help the reader's understanding of *theion* is "a *theion* is to divine energy and divine light as a *photon* is to physical energy and physical light." Just as a *photon* is a force-carrying, massless particle in the physically-observable universe, so a *theion* is a force-carrying, massless particle in the spiritually-observable universe. In the physically-observable universe, all subatomic particles are actually dense packets of bound energy derived from the energy originally released from a minuscule number of *theions* at the time of *the Luciferian Fall.* And the release of massive energy from the fission of just one *theion* is analogous to the release of massive energy from the fission of just one atom except on a much grander scale and more powerful level.

One measure of the utility of the photon to *theion* analogy is in the capacity of the units to self-replicate or not. Because photons are not able to self-replicate and *theions* are able to self-replicate, the photon to *theion* analogy is less than perfect. However, it is still a useful analogy, and conceptualizing *theions* provides a useful tool for those who might like to compare spiritual energy to physical energy.

In brief, matter and physical energy (in contrast to Spirit and spiritual energy) were initially formed at the time of the introduction of iniquity into a portion of *the Whole Universe* with *the Luciferian Fall.* Previous to this introduction of iniquity into *the Whole Universe,* there was no temporality, no relative space-time, no matter, and no physical energy.

From one valid metaphysical perspective, the introduction of iniquity into a portion of *the Whole Universe* resulted in an evagination from eternity that formed the originally chaotic physically-observable universe with infinite space beyond its fringes (i.e., its cosmic horizon). From a second valid metaphysical perspective, a parallel universe was made that cast a shadow, or pall, over the Lord God Almighty's original creation. And, from a third valid metaphysical perspective, whereas *Paradise* was all that existed prior to *the Luciferian Fall,* there now existed something *outside of Paradise* (i.e., in the *elsewhere* of Paradise). Each of these three perspectives constitutes a thread in a metascopic view (i.e., a metaphysically holistic perspective) of the first aspect of *displacement* (the origin of matter and physical energy in relative space-time).

For the sake of clarification at this juncture, *Paradise* is synonymous with a state of immortality and *outside of Paradise* is synonymous with a state of mortality — which are the only two possible states of being that currently exist within *the Whole Universe.* These two states are further explained in the next numbered section (3.2.2.2).

In the Bible, *the Luciferian Fall* is represented by what is *not* stated in between verses 1 and 2 of Genesis, Chapter One:

> {1} In the beginning God created the heaven and the earth.
> {2} And the earth was without form, and void; and darkness *was* upon the face of the deep. And the Spirit of God moved upon the face of the waters. [italics mine]

The specific tense of the Hebrew verb hä·yä [H1961], translated as the italicized "was" in verse 2, can also be translated as "became," implying that at least one aspect of the Creator-Savior's original "heaven and earth" (verse 1) *became* formless and void (i.e., chaotic, confusive, and disordered) after its creation upon the introduction of iniquity, or spiritual darkness, into the Creator-Savior's spiritually-observable universe. Despite the formation of this state of chaos, confusion, and disorder, we see from reading the second sentence in Genesis 1:2 that the Creator-Savior moved upon this newly-formed medium to bring order out of its unique form of entropy. (*Entropy* represents the degree of disorder in a system.)

Biblically speaking, *iniquity* is sometimes referred to as "the shadow of turning" because, in simple terms: (1) *iniquity* is the spiritual product of "turning away" from the Lord God Almighty by refusing to obey His Will; and (2) *iniquity* casts a shadow, or pall, over the Creator-Savior's original creation, hiding it from fallen beings who have banished themselves to mortality and are, subsequently, held hostage by the evil they helped to bring into existence.

3.2.2.2 The Origin of Eternal Souls in Corporeality

The second aspect of *displacement theory* is the origin of eternal souls in corporeality. This aspect includes the supposition that created eternal souls fell from immortality to mortality — or from a state of immortal life in *Paradise* to a state of mortal death *outside of Paradise* — at the moment that they turned away from the Creator-Savior by going against His Will. For the paradigm of *intelligent evolution,* the turning away of these eternal souls is represented in Genesis by the Adamic Fall. Their *turning* represents the introduction of iniquity into their eternal souls as well as their souls' immediate displacement from a state of immortality to a state of mortality.

Although *Adam* is (1) the name of a historic human being in the Bible, *Adam* is also (2) the name for a corporate structure of originally unfallen beings as well as (3) the name for the race of fallen souls who have collectively come to reside in corporeality as individual human beings. To be sure, the shadows cast by the fallen Lucifer (i.e., Satan, or eternal Adversary, of the Creator-Savior) played an important role in the temptation of unfallen Adamic souls to disobey the Creator-Savior's Will. That is why fallen Adamic souls have an opportunity for salvation and Lucifer and his fallen angels do not. Lucifer and his fallen angels were not influenced to fall by anything external. In contrast, fallen Adamic souls were influenced to fall by something external (i.e., their temptation by Lucifer).

Within the paradigm of *intelligent evolution, mortality* is not synonymous with either *physical death* or *corporeality.* Although a state of mortality (i.e., that-which-is-outside-of-Paradise) includes a corporeal condition of being and physical death because of that condition, it also includes an incorporeal condition of being where *unsaved* fallen eternal souls either await their turn to be born into corporeality or, because they have proven themselves to be beyond redemption, await their final judgment at the end of relative space-time. In contrast, *saved* fallen eternal souls have been restored to immortality regardless of where their souls are — either (1) still in corporeality because their human bodies have not yet expired or (2) returned to *Paradise* as "the dead in Christ" because their human bodies have already expired. In other words, regardless of where they are (either in *corporeality* or in *incorporeality),* (1) unsaved fallen eternal souls are *mortals,* and (2) saved fallen eternal souls are *immortals.* To shed further light on this sense of the word *immortals,* the Hebrew word el·ō·hēm' [H430], translated as "gods" in Psalm 82:6, could have been translated as "immortals" in that verse — meaning, the "children of the Most High God" are *immortals* (but not as objects of worship). Albeit perverted, the concept of *immortals* from the Creator-Savior's spiritually-observable universe was retained in Greek, Roman, and other ancient mythologies as the basis for anthropomorphized gods and goddesses — which is to say, immortal beings with human personality traits and characteristics.

3.2.3 The Origin of the Supreme Being's Tri-Unity

The Supreme Being has always been One and will always continue to be One. However, at the time of the beginning of *intelligent evolution* — as recounted in Chapter One of Genesis — the Supreme Being was partitioned into: (1) *the Lord God,* (2) *the Spoken Word (the Logos of God),* and (3) *the Spirit.* Alternate titles for these three partitions include: *God the Father, God the Son,* and *God the Holy Spirit.* People who think that the Supreme Being's tri-unity is representative of three different deities are incorrect; they are using the wrong mathematical model for their conceptualization. Instead of $1 + 1 + 1 = 3$, the correct mathematical model for the tri-unity of God is $1 \times 1 \times 1 = 1^3$, or one raised to the third power (1^3). No member of this tri-unity operates independently of the other two. All three, in fact, are One.

The earliest identification in the Bible of the tri-unity of the Supreme Being is found in Genesis, Chapter One, verses 1 through 3:

> {1} In the beginning God *[God the Father]* created the heaven and the earth. {2} And the earth became formless and void; and darkness was upon the face of the deep. And the Spirit of God *[God the Holy Spirit]* moved upon the face of the waters. {3} And God *said [said* representing *God's Spoken Word, the Logos of God,* or *God the Son]:* "Let there be light: and there was light." [brackets and italics mine]

That *the Spoken Word* in Genesis 1:3 is *the Son of God* is confirmed in the Gospel According to John, Chapter One, verses 1 and 14:

> {1} In the beginning was the Word *[the Logos]*, and the Word *[the Logos]* was with God, and the Word *[the Logos]* was God. {14} And the Word *[the Logos]* was made flesh, and dwelt among us, (and we beheld his glory, the glory of the

> *only-begotten* of the Father) full of grace and truth. [brackets mine]

Christ Jesus is also called *the Word of God (the Logos of God)* in Revelation 19:13:

> And he [Christ Jesus] was clothed with a vesture dipped in blood: and his name is called *the Word of God [the Logos of God]*. [brackets and italics mine]

The Creator-Savior had partitioned Himself in order to effect His Plan of Salvation for Adamic souls who would fall from immortality to mortality. The Creator-Savior wanted to retrieve all fallen eternal souls who would eventually repent of their waywardness in exalting themselves and, instead, return to exalting Him.

At the time of the end (after the millennial reign of Christ Jesus on Earth), when all that is to be restored to the Creator-Savior has been restored, *God the Father* will then infuse the Totality of His Being (i.e., *His Fiery Presence*) into the "all" that He has placed under the feet of *God the Son*. At that time, there will no longer be partitions of the Supreme Being or separation of the Supreme Being from His created souls because the Creator-Savior will then be *All-in-all*. Although the Creator-Savior is "All," He is not technically "in all" until the time that Christ Jesus' millennial rule on Earth is over. This infusion and reunification are attested to in 1 Corinthians, Chapter Fifteen, verses 24 through 28:

> {24} Then comes the end, when he *[God the Son]* shall have delivered up the kingdom to God, even the Father — when he *[God the Son]* shall have put down all rule and all authority and power. {25} For he *[God the Son]* must reign until He *[God the Father]* has put all enemies under his *[God the Son's]* feet. {26} The last enemy that shall be destroyed is death. {27} For He

> *[God the Father]* has put all things under his *[God the Son's]* feet. But when it says all things are put under him *[God the Son]*, it is clear that He *[God the Father]* is exempted who put all things under him *[God the Son]*. {28} And when all things shall be subdued unto him *[God the Son]*, then shall the Son also himself be subject unto Him *[God the Father]* who put all things under him *[God the Son]*, that God may be all in all. [brackets mine]

To be sure, Christ Jesus ("God the Son") already has all authority and all power in Heaven and on Earth *(Matthew 28:27 and Ephesians 1:22),* but not every enemy has been finally conquered yet, or "subdued unto him" *(1 Corinthians 15:28). For example,* the end-time Antichrist has not yet been overcome. And *death* — not just physical death but the state of mortality itself — remains to be expunged.

Scripture teaches that the final end-time Antichrist will not be thrown into the Lake of Fire until the time of Christ Jesus' return to Earth *(Revelation 19:20)*. Scripture also teaches that *death,* or the state of mortality itself, will not be conquered until the end of the millennium of peace, when *death* — along with Hades (the current holding tank for unsaved fallen souls) — will be thrown into the Lake of Fire at the time of the Great White Throne Judgment *(1 Corinthians 15:26; Revelation 20:11-14)*, during which time each remaining soul will either be assigned to a state of eternal redemption or a state of eternal damnation.

At this juncture, it is important to note that the Creator-Savior, His Created, and His Re-created are fully united only when *God the Father* becomes *All-in-all.*

3.2.4 The Cult of Physical Entropy

Entropy is the loss of energy resulting in a gradual change from order and non-randomness toward disorder and randomness in a physical system. Although entropy is often represented in schemata with objects or pictures of objects that are visible to the naked eye, *entropy* is actually a microscopic/submicroscopic phenomenon that impacts physical systems at colloidal, molecular, atomic, and subatomic dimensional levels: *Entropy* is not a phenomenon that directly impacts the movements or orbits of celestial bodies like asteroids, comets, moons, planets, and stars.

The Cult of Physical Entropy refers to the godless notion that, once energy is lost from a physical system (especially in the form of heat), it can never be regained or replaced. This notion is *godless* because it presupposes that the Creator-Savior is unable to infuse new energy into the physically-observable universe. People who subscribe to the law of physical entropy to predict all final and ultimate changes in the physically-observable universe are considered members of a cult by the present author because: (1) they reject creation by a hyperpersonal[27] Supreme Being; (2) they hold viewpoints like *intelligent evolution* in contempt and with disdain; and 3) they try to constrain others to accept their godless paradigm as the only viable paradigm. Regardless, the law of physical entropy neither proves nor disproves *intelligent evolution*.

Despite the demonstrable existence of physical entropy, the Creator-Savior continues to sustain the physically-observable universe and continues to infuse His energy into it in order to maintain order in a steady state as well as to correct, right, and heal

[27] ***Hyperpersonal*** **here is not referring to phenomena associated with cybertalk. Instead, *hyperpersonal* is referring to relationships and communications that are spiritually beyond, or metaphysically higher than, corporeal face-to-face interactions.**

all types of physical imbalances. (That "all types" are corrected, righted, and healed does not mean that all imbalances will be restored to equipoise; it simply reaffirms that all options remain available to the Creator-Savior.) The Creator-Savior is responsible for any and all negative entropy (i.e., *negentropy* or entropy deficit) in the physically-observable universe.

Interestingly, because simultaneous *entropy* in multiple adjacent systems can contribute to complexification in the intermingling interactions of those adjacent systems, physical entropy is one method, process, or means that the Creator-Savior used in the origin of life on Earth during the time referred to by the present author as *chemicalization* (see Section 3.2.6).

3.2.5 About the Origin of Biological Life on the Planet Earth

Whether one concludes that the Creator-Savior originated biological life on the planet Earth *ex nihilo* or *de novo* ultimately depends on how one views matter. If you view matter as metaphysically *nothing* ("no thing") when compared to the glories of the spiritually-observable universe, then your conclusions may well include that the Creator-Savior originated biological life on the planet Earth *ex nihilo* (i.e., *out of nothing).* Or, if you view matter as originally devoid of life and that the Creator-Savior instilled, or breathed, life into the matter that He aggregated and formed as physical organisms, then your conclusions may include that the Creator-Savior originated biological life on the planet Earth either (1) *ex nihilo* by *creating* biological life from non-life or (2) *de novo* by *making* biological life out of previously non-aggregated, or unformed, matter. To be sure, as *ex nihilo* and *de novo* are used in the last sentence, there can be an overlap of meaning for the two phrases. Thus, one may even conclude that the Creator-Savior *created* all elements in the universe *ex nihilo* during the events of *the Big Bang* (i.e., relatively early in the inflationary epoch) and that He later *made* all biological life *de novo* out of those elements in their various combinations.

Technically, for the Supreme Being, "creating" is when He forms something from nothing, and "making" is when He takes from what already exists to form something new. The major distinction between *ex nihilo* and *de novo* here reduces to how the original forms of biological life came to be through the Creator-Savior's direction, control, and intervention. Thus, depending on how you look at matter, forming the first living cell from non-living components is either *ex nihilo* or *de novo.* (However, cloning a second cell from a first cell — or a biological Eve from a biological Adam — would always be *de novo.*)

The age-long debate concerning the concepts behind *ex nihilo* and *de novo* is underscored by what is presented under the heading *Creation* in the *1906 Jewish Encyclopedia:*

> Most Jewish philosophers find in [the Hebrew word] bä·rä' [H1254] (Genesis 1:1) creation *ex nihilo.* The etymological meaning of the verb bä·rä' [H1254], however, is "to cut out and put into shape," and thus presupposes the use of material. [Brackets are mine with the numbers from Strong's Hebrew Lexicon.]
>
> *Jewish Encyclopedia, Volume 4, p. 336*

Regardless of connotations assumed or perspectives derived from the verbiage used in the Bible or elsewhere, all biological life on Earth was created by the Creator-Savior through *intelligent evolution,* whose processes include: (1) chemicalization of precursors necessary for biological life, (2) origination of all biological life, (3) natural variations within each species, and (4) diversification of biological life through successive speciation.

3.2.6 Chemicalization of Precursors Necessary for Biological Life

For the present author, *chemicalization* is the God-driven formation of precursors necessary for biological life. As such, *chemicalization* includes the first few steps in the crystallization of the Creator-Savior's thinking in the origin of biological life on the planet Earth. Although *chemicalization* took place over billions of years, what occurs in relative space-time is neither slow nor fast to the Creator-Savior because He resides in eternity.

Chemicalization actually began during *the Big Bang,* when the Creator-Savior organized discrete bundles of energy — including quarks, gluons, and electrons — to form the protons and neutrons that eventually aggregated together to constitute the various nuclei of atoms that exist as ordinary matter within the physically-observable universe. However, *chemicalization* not only includes the nucleosynthesis that occurred during *the Big Bang* but also the nucleosynthesis that continues to occur today in stellar activities (the latter nucleosynthesis implying that, at least in this way, physical creation is not really over).

The atoms of elements resulting from nucleosynthesis each have their own individual physical and chemical properties that enable them to form weak and/or strong energy bonds with other atoms, which bonds result in aggregates of atoms, ions (i.e., electrically-charged atoms), compounds, and/or molecules — each capable of interacting with other individual atoms, ions, compounds, and/or molecules. Indeed, on their own, some aggregates of atoms, ions, compounds, and molecules have tendencies toward continued bonding, aggregating, cohering, polymerizing, vulcanizing, separating, layering, boundary-making, enveloping, dissociating, interacting, and integrating with other aggregates to form supramolecular assemblies. Some supramolecular assemblies can even: (1) move through twisting, rotating, folding, and sliding; and (2) partition themselves off from the microenvironments in which they are found. However, it should be noted that the movement and partitioning achieved by such supramolecular assemblies do not constitute life — although self-control, self-organization, and self-containment are properties associated with biological life.

To be sure, *chemicalization* depends on the impetus given by the Creator-Savior in the conjoining of atoms, ions, compounds, and/or molecules in the primordial sea of the early Earth in order to form organic building blocks and supramolecular assemblies important to the origin of the earliest biological life.

Although some modern scientists have been able to duplicate conditions thought to have existed on Earth when biological life originated, they have only been successful in producing some of the precursors necessary for biological life: They have not been successful in producing biological life itself. Indeed, modern science will never be able to create biological life because modern science cannot replicate or invoke the Supraconsciousness of the Author of all life that gave impetus to the inanimate atoms, ions, compounds, and molecules that He assembled together and to which He alone imparted life. It is the Supraconsciousness of the Creator-Savior that provided the necessary driving force, impetus, and spark for the origin of the first biological life on Earth, regardless of whether lightning or other electrical sparks played a role or not.

Although the present author has a personal aversion to what is generally conveyed by the word *pantheism,* he admits that what he has written — concerning (1) nucleosynthesis and (2) the properties of some aggregates of atoms, ions, compounds, and molecules permitting them to organize and move on their own (leading up to the origin of biological life but not responsible for the origination itself) — represents a type of immanence that embodies a transcendent pantheism imparted to matter by the Creator-Savior at the time of *the Big Bang* and continuing after it. However, because it is recorded in Genesis that the Creator-Savior rested (i.e., stopped) after His six sequential steps of creation, the origination of biological life from non-life is not a recurring phenomenon on Earth regardless of current or future physical and environmental conditions that may occur naturally or artificially. *For example,* the laboratory injection of DNA into an enucleated cell does not constitute creating; *that* constitutes reassembling. Although modern science is increasingly capable of such reassembling, it is the Creator-Savior alone who is capable of creating biological life out of inanimate matter.

The *chemicalization* of precursors necessary for biological life depended on *directed chance. Directed chance* is the deific Force of the Creator-Savior that guided design and purpose in the formation of chemical precursors necessary for the origin as well as the diversification of biological life. Although no directed outcomes of *chemicalization* were accidental, some associated outcomes were incidental to, or coincidental in, the formation and maintenance of these precursors. Make no mistake, *directed chance* is an effect of the Supraconsciousness of the Creator-Savior; it is not Supraconsciousness-in-itself. (Do not confuse an effect of Supraconsciousness with Supraconsciousness-in-itself.)

Directed chance not only played an important part in the first appearance of biological life on Earth, *directed chance* continued to play an important role in the diversification of biological life through combinations and permutations of variations and adaptations that resulted in speciation. *Directed chance* is so much a part of *intelligent evolution* that *intelligent evolution* could also be called *directed evolution. Directed chance* articulates the major principle in *intelligent evolution* that the Creator-Savior's intended outcomes, not physical causes, guided all cosmic evolution, biological evolution, and consciousness evolution for the first purpose and final cause of members of *Homo sapiens* serving as suitable host organisms for fallen eternal souls. Indeed, the Creator-Savior's *directed chance* ensured the intelligent evolution, origin, survivability, thrivability, and sustainability of the species *Homo sapiens.*

It is important to add here that, although de Chardin used the phrase *directed chance* once in *The Phenomenon of Man* (on page 110), it was used by de Chardin in the sense of "groping," experimenting, and trying out different variations in order for consciousness to discover outcomes *a posteriori* rather than in the sense that the present author uses *directed chance* — where specific outcomes are already known by the Creator-Savior *a priori* before He directs and guides them to happen.

3.2.7 Tandem Creations

Today, there are tandem creations of the Creator-Savior. The Creator-Savior's two creations include: (1) the one that took place in the spiritually-observable universe and its eternity (in the *here* and *now*); and (2) the one that took place in the physically-observable universe and its temporality (i.e., in the relativity of space and time). Before the beginning of relative space-time there was only one creation. That first creation in the spiritually-observable universe comprised the entirety of creation and constituted *the Whole Universe*. Today, however, *the Whole Universe* is composed of the physically-observable universe in addition to the spiritually-observable universe. Today, *the Whole Universe* is no longer filled only by the spiritually-observable universe.

What we now know as the physically-observable universe is a result of the Creator-Savior's bringing order to the explosion of chaotic energy that ensued upon the introduction of iniquity into *the Whole Universe*. Iniquity was introduced into an infinitesimal portion of *the Whole Universe* at the first instant of Lucifer's presumption that he could become greater than the Creator-Savior (needless to say, he cannot). The introduction of Lucifer's newly-formed iniquity into the spiritually-observable universe altered a minuscule number of *theions* of the Creator-Savior's divine energy, resulting in: (1) the chaotic energy seen during the first nanosecond of *the Big Bang* as well as (2) the outpocketing from eternity of what is now known as temporality (review Figure One on page I-23). Without pause, the Creator-Savior intervened to bring order to this chaotic energy by creating the heretofore unknown substance we call *matter*. The creation of this new substance began with the Creator-Savior's nucleosynthesis of the more than one hundred elements of ordinary matter (i.e., fundamental, or baryonic, matter) that now exist.

It is through the formation of atomic nuclei that the Creator-Savior first began to bring order out of the iniquity-induced energy chaos. The Creator-Savior brought additional order to this newly created matter through its cosmic evolution, biological evolution, and consciousness evolution with the driving forces of *directed chance* and *chemicalization* — all of which is collectively referred to by the present author as *intelligent evolution*. In one way, despite its order by the Creator-Savior, ordinary matter itself is the physically visible sign of iniquity. However, matter is neither to be feared nor hated. Matter itself is not evil. Matter is merely the substance, or essence, of the physically-observable universe — just as Spirit is the substance, or essence, of the spiritually-observable universe.

Because the Creator-Savior's first creation was spiritual, created beings with eternal souls in the spiritually-observable universe each had a somatic identity before the Adamic Fall that was spiritual in nature and not physical in nature. In his various metaphysical works, the present author has described this original somatic identity as *astral gelatinous™*.

Note: The phrase *astral gelatinous™* was coined by the present author and first copyrighted in the 2011 edition of his work entitled *Divine Metaphysics of Human Anatomy* (United States Copyright Office TXu001788674).

The phrase *astral gelatinous™* describes a substance that predominantly has spiritual qualities similar to the created substance of unfallen angels. This substance may also take on physical qualities depending on the dimensionality in which it is found or into which it is projected. *For example,* when some angels step into the physical realm (i.e., *push* themselves into relative space-time), they voluntarily take on human form and appear to be human even though they did not originate from, or in, a hominin life form. This is exemplified by the angels who first visited Abraham and, later, Lot in the city of Sodom — which visitations are recorded in Chapters Eighteen and Nineteen of the Book of Genesis. At one time, certain angels even stepped into physicality in order to mate with human beings. Their sexual interaction is recorded in Genesis 6:1-4 as having taken place between "the sons of God" and "the daughters of men." The giants, or *nephilim* (i.e., "fallen ones"), mentioned in Genesis 6:4 and Numbers 13:33 of the Holy Bible were the offspring of these unnatural sexual liaisons. The Holy Bible is clear that the spiritual beings who mated with human beings are now relegated to the *bottomless pit of the Abyss in Hades,* awaiting the Creator-Savior's final judgment and condemnation for their transgression (see verse 6 of the Epistle of Jude).

The unfallen creation that originally reflected the Creator-Savior's complete image and perfect likeness was *astral gelatinous™* in nature (i.e., *in essence*). As a result of the Adamic Fall, the *astral gelatinous™* substance of immortal beings, originally created in the complete image and perfect likeness of the Creator-Savior, manifested as living physical substance (i.e., protoplasm). Consequently, the various cells, tissues, organs, and organ systems of the modern hominin appeared, becoming mere representations, vestiges, remnants, and "fossilized impressions" of what they used to be. In other words, as the fallen Adam and Eve were expelled from the Garden of Eden, they materialized as human beings and appeared alongside other human beings who had evolved on Earth without souls. *(For example,* Neanderthal men and women did not possess eternal souls.)

As mentioned earlier, at the return of Christ Jesus, all who are joint heirs with him shall receive their new somatic identities. These new somatic identities not only resemble the body of the ascended Christ Jesus but also are composed of the same *astral gelatinous™* substance that constituted the somatic identities of immortal beings before the Adamic Fall.

In summary, *astral gelatinous™* substance is the spiritual substance of immortal beings. It has spiritually-translucent, spiritually-luminescent, and spiritually-iridescent qualities that reflect the *glory,* or spiritual light, of the Creator-Savior. *Astral gelatinous™* substance is spiritually light, airy, and shimmering.

For the sake of clarity, gender and sexual identity do not exist in an *astral gelatinous™* condition of being. Instead, beneficial mental and emotional characteristics associated with each gender and sexual identity on Earth are fused together for each entity in Heaven. In other words, there are no males, females, hermaphrodites, or intersexuals in Heaven. All beings in Heaven are spiritual, not physical.

3.2.8 Concerning Sexual Dimorphism in Human Beings

Metaphysically speaking, maleness and femaleness are not only abstract constructs but also artificial concepts foreign to the nature, or essence, of the beings who people the Creator-Savior's spiritually-observable universe. For those familiar with common terminology in biology, eternal souls in Heaven are neither *monoecious* (monecious) nor *dioecious* (diecious). However, in the most highly evolved biological forms on Earth, sexual dimorphism does exist. Especially for vertebrates, sexual dimorphism is the separation of male and female reproductive organs into two different body forms. Thus, human beings are *dioecious* — meaning, their reproductive organs are found in "two separate households." All species with sexual dimorphism have a survival advantage because a greater number of genetic variations exist in their progeny and, as a result, increased survivability, increased sustainability, and increased thrivability are imparted to the entire species. As a side note, the ambiguity of internal and external genitalia in human beings during their early embryonic development metaphysically allude to the genderless nature that eternal souls possessed in their original *astral gelatinous™* somatic forms.

At the time that fallen eternal souls first appeared in corporeality in the form of *Homo sapiens* as a result of their (1) newly-developed iniquity and (2) concomitant expulsion from *Eden* (i.e., *Paradise* or *Heaven),* there already existed on the planet Earth an evolved race of *Homo sapiens* without eternal souls. The Creator-Savior's *intelligent evolution* of *Homo sapiens* without eternal souls on the planet Earth occurred because the Creator-Savior: (1) foreknew that there would be an Adamic Fall of spiritual beings from Heaven to Earth as well as (2) foreknew what their physical appearance and physiology would be. The Creator-Savior foreknew what their physical appearance and physiology would be because He understood that, under His guidance and direction, their physical appearance and physiology would represent a concretioning of the *astral gelatinous™* somatic forms and functions that He had created in the spiritually-observable universe. Because of His foreknowledge and understanding, the Creator-Savior created biological life in His second creation that would be comparable to, and compatible with, the reproductive capabilities of fallen Adamic beings — ultimately resulting in hybridized offspring that could also house fallen eternal souls. (The children of Cain and Seth were the first such hybridized offspring.) It was also predetermined by the Creator-Savior that, in such human housing, fallen eternal souls would be given opportunities for: (1) salvation by accepting Christ Jesus as their Savior; and (2) re-sanctification by allowing the Creator-Savior's Holy Spirit to, once again, freely permeate their souls and inform their *being* of the Creator-Savior's Will for them individually, collectively, and corporately.

3.3 Intellectual Proofs for Intelligent Evolution

Using evolution, the God of the Holy Bible created and diversified all biological life on Earth. Because the Creator-Savior caused all biological life to evolve, the overall process is called *intelligent evolution*. As stated previously, *intelligent evolution* includes these three general processes: (1) cosmic evolution, (2) biological evolution, and (3) consciousness evolution. Because *directed chance* and *chemicalization* played roles in all three general processes, they are not listed here as additional processes. And, to be sure, even though the three general processes are listed separately, all three of them are interdependent for the purpose of *intelligent evolution*.

The importance of *intelligent evolution* is that it helps students of Christian metaphysics to blend, harmonize, meld, and synthesize the two most important paradigms on Earth. The first most important paradigm on Earth is salvation through the shed blood of Christ Jesus, the *only-begotten* Son of the Creator-Savior. The second most important paradigm on Earth is the evolution of biological life, including its origin and diversification. Unfortunately, many people who have accepted the paradigm of salvation have rejected the paradigm of evolution because they have been taught that evolution is contradictory and in opposition to a belief in Christ Jesus. And many people who have accepted the paradigm of evolution have rejected the paradigm of salvation because they have been taught that salvation through Christ Jesus is mythological and/or not provable scientifically. (The present author maintains that experiencing salvation through Christ Jesus is part of one's journey that is synchronously personal, interpersonal, and hyperpersonal.)

For the sake of clarification, many people (if not most people) who accept the paradigm of salvation through Christ Jesus presuppose that the God of the Holy Bible created the Earth and all types of biological life within six twenty-four hour periods of time (such supposition often referred to as *creationism).* By adopting this view, creationists think that, because they are not challenging what they believe is the literal truth contained within the Bible, they are being respectful to the Creator-Savior. Other saved people believe that creationism and evolutionary theory can coexist without the need for blending, harmonizing, melding, or synthesizing them. In other words, some saved people actually hold as true what they think are two opposing or contradictory views without acknowledging it to be a problem. Regardless of their acknowledgement or not, a problem does exist because holding these two views as opposing or contradictory at the same time that one holds both views as true diminishes one's ability to witness effectively to many educated unsaved people: *For example,* they are unable to explain to academically-oriented people how the physically-observable universe came to be in relationship to the God of the Holy Bible. And, without a reasonable explanation to offer about cosmic evolution, biological evolution, and consciousness evolution, many educated unsaved people are unwilling to make the cognitive leap from the message of six twenty-four days of creation to the message of salvation through the shed blood of Jesus Christ.

Because this section (3.3) has been developed as an evolving list of intellectual proofs for *intelligent evolution,* students of *intelligent evolution* are encouraged to: (1) discuss this list with like-minded people; (2) reshape the existing individual proofs to enhance the list's clarity and accuracy; and (3) add proofs to this list.

Because *intelligent evolution* is a metaphysical tool, its individual proofs need to make sense to students of *intelligent evolution* at the same time that the students are reminded that every aspect of each individual proof cannot be: (1) attested to via experimentation, (2) provable *a posteriori,* or (3) testable through physical experiences and calculations (i.e., quantitative analysis). Indeed, some intellectual proofs are only analyzable in one's imagination through inner dialogue, metaphysical understanding, and pure reason.

In order to devise proofs for *intelligent evolution* that are based on metaphysical understanding and pure reason, one must look for common ground and compatibility between creationism and evolution and capitalize on their common ground and compatibility by constructing intellectual proofs that are intelligible to inquiring minds that have been properly nurtured, trained, educated, and disciplined. Of course, without proper nurturance, training, education, and discipline, some minds will never be reached to understand the paradigm of *intelligent evolution*. Without a mentally and emotionally healthy environment during their formative years, the minds of some children easily regress, or revert, to a Neanderthal mentality, emotionality, and brutality that may remain with them for the duration of their lives. In such cases, rather than focus on *intelligent evolution,* the present author believes that the primary responsibility of saved people is: (1) to clearly share with them the message of salvation through the Biblical Jesus (in contrast to, *for example,* a fictitious New Age Jesus, an Avatar Jesus, a HinduChrist, or a Chrislam Jesus) and (2) to hope for change in them by the Creator-Savior's Holy Spirit, the real teacher of all truth. Without a belief in Christ Jesus as one's personal Savior, and without the Creator-Savior's Holy Spirit residing within us, any one of us can easily regress, or revert, to animal instincts, brutality, cunningness, pugilism, and vulgarity — even if one has a high IQ, possesses great wealth, and has employment in an area with significant prestige and responsibility.

3.3.1 The Immensity of the Physically-Observable Universe

The present author has chosen the immensity of the physically-observable universe as a primary intellectual proof for *intelligent evolution* because the night sky evokes awe in its observers regardless of the major paradigm to which they subscribe. Indeed, awe is a first step in exalting the Creator-Savior. Unfortunately, due to light pollution and smog, few people today have a spectacular view of the starry night. At best, they see an evening or morning planet or two, the moon, and a few of the closest and brightest stars. Seeing so few night objects, one can easily become jaded and blasé about the night sky because there is no good draw or real incentive to look up at it.

Therefore, in order to use the magnitude of the starry sky and its significance as an intellectual proof for *intelligent evolution,* one must first visit a planetarium, an astronomical observatory with accessible telescopic views, the top of a rural mountain, a remote island, or a region within a country that provides panoramic night views *(for example,* from a hut on the island of Bora Bora or an arctic glass igloo in Rovaniemi of Lapland, the northernmost region of Finland). If a teacher cannot afford to take his or her class on such an extravagant field trip, then the teacher must use projected slides and digital images to show his or her students impressive views of the night sky to study and discuss.

Impressive views of the night sky should be used to facilitate discussions about (1) stars, (2) the distances of stars from one another, (3) the distances of stars nearest to the Earth, (4) what a light-year is, (5) how many miles one light-year represents, (6) the movement of stars away from one another, (7) the expansion of the physically-observable universe, (8) what the expansion suggests concerning a "center" of the physically-observable universe, (9) whether the physically-observable universe has a flat or spherical shape, and, eventually, (10) calculating how long it would take for human beings to travel from the Earth to a close prominent star like Alpha Centauri (A or B) or Capella (A or B).

Although Capella (consisting of Capella A and Capella B) is not the closest or the brightest of stars, it can easily be seen in the night sky of the Northern Hemisphere at certain times of the year even from a city subjected to light pollution and smog. (If Capella is not visible where you live, then choose another star that is visible.) The present author especially likes Capella because it flashes rather than just twinkles (stars usually twinkle, planets don't). Capella is approximately 42 light-years away. Therefore, because one light-year is approximately six trillion miles, Capella is approximately 250 trillion miles away from Earth. If human beings boarded a spacecraft that travelled 1,000 miles per hour, then they would be travelling 24,000 miles per day and approximately 9,000,000 miles per year. At that velocity, it would take them approximately 28 million years to visit Capella!

A discussion about the distances between stars or between stars and the planet Earth, and how long it would take us to travel between stars or between a star and the planet Earth, should eventually lead students to think about duration of travel time. A discussion about distance and duration can help students to conclude that the stars were not created in just six twenty-four hour periods of time. It should be emphasized to students that they are not insulting, blaspheming, or ridiculing the Creator-Savior to make such an observation or think such thoughts.

Additionally, a discussion about the physically-observable universe having a center, and that all stars are moving farther away from that center, can help students comprehend that the physically-observable universe originated from a central point during *the Big Bang* and that all stars are continuing to move away from that central point — causing the expansion of the universe away from its origin. Because the outer fringes of the physically-observable universe are somewhere between 46.5 and 48.0 billion light-years away from that central point (constituting its radius), then the diameter of the entire physically-observable universe is between 93 and 96 billion light-years wide. On a chalk board or white board, teachers should multiply 96 billion light-years by 6 trillion miles per light-year, showing all zeroes in their calculations, to impress upon the students the immensity of the physically-observable universe. Also, teachers should make the following important point: If the

Creator-Savior hung all stars in the physically-observable universe within the first four days of creation, there would be no need for Him to cause them to continually move away from a central point. To be sure, each star could have remained relatively stationary in relationship to all other stars without negatively impacting the existence of biological life on Earth.

(1) A discussion about cosmic durations and movements should interest inquiring Christian minds in a discussion about cosmic evolution. And (2) a discussion about their awe of the starry night should interest scientifically-minded people in a discussion about creation and whether one should worship and exalt the created objects in the physically-observable universe or the Creator of those objects.

The next section (3.3.2) may be helpful in facilitating a discussion about the origin of the physically-observable universe from both physical and metaphysical standpoints.

3.3.2 Metaphysical and Physical Origins of Matter and Temporality

Throughout the physically-observable universe, the atoms of all elements are composed of the same dense packets, or discrete bundles, of *physical energy* known as elementary, or fundamental, particles — each of which is subatomic (i.e., smaller than an atom). Here, the phrase *physical energy* is used in contradistinction to the synonymous phrases *eternal energy, divine energy,* and *spiritual energy* — all of which represent the type of energy found in the spiritually-observable universe.

That all subatomic particles represent bound energy alludes to their formation as an explosive aftereffect of *eternal energy* first colliding with *iniquity* — which explosive collision: (1) caused a momentary tear, rip, hole, gap, or rift in the spiritually-observable universe; and (2) created a vacuum that sucked the newly-formed chaotic energy in a direction away from the central point of its formation (that vacuum is what we think of today as *outer space).*

Any and all atomic order that exists in the physically-observable universe was imposed on *physical energy* by the Creator-Savior in His immediate response to the formlessness, void, and vacuum that resulted from the explosive collision of *eternal energy* with *iniquity*. That the atoms of all elements and their isotopes can be arranged sequentially — based on the number of protons and neutrons that the nucleus of each atom has — demonstrates the atomic order that the Creator-Savior brought to the formlessness, void, vacuum, and chaotic energy caused by iniquity's explosive interaction with eternal energy. To be sure, the numbers and combinations of subatomic particles vary from the atoms of one element to another, but the kinds of subatomic particles remain constant. In other words, the sequenced arrangement of elements on the Periodic Table represents one form of order in the early physically-observable universe that was directed, determined, and imposed — as well as continually maintained — by the Creator-Savior. The Creator-Savior transformed the formlessness, void, vacuum, and chaotic energy into order because it pleased Him to do so, and because it was part of His plan for the eventual salvation and restoration of fallen eternal souls.

Just as Einstein's equation $E = mc^2$ best represents the interconvertibility of matter and *physical energy*, so does the following metaphysical reaction equation best represent the formation of matter and temporality from the contact of *eternal energy* with *iniquity:*

$$E^{o} + \bar{E}^{o} \rightarrow m + e^{o} + t = T = \Sigma D$$

where E^{o} is *eternal energy*, $\bar{E}^{o}$ is iniquity (anti-*eternal energy*),
m is physical mass, e^{o} is physical energy, t is physical time, T is Temporality,
and ΣD is the sum of all dimensions in the physically observable universe

The conclusion here is that temporality — where physical mass, physical energy, and physical time (i.e., relative time) exist — is the sum of all dimensions in relative space-time. As a result, *temporality* is equivalent to physical *dimensionality*. (Please remember that eternity is dimensionless.) Corollaries include: (1) eternity does not house physical mass, physical energy, or physical time; (2) where physical mass and physical energy do not exist, physical time does not exist; (3) where physical time does not exist, physical mass and physical energy do not exist; (4) when eternity is divided (i.e., interrupted) by physical time, temporality results; (5) where physical time exists, physical mass and physical energy also exist; (6) where physical mass and physical energy exist, physical time also exists; (7) temporality houses physical time, physical mass, and physical energy; and, finally, (8) dimensionality houses physical mass, physical energy, and physical time as metaphysical functions of iniquity.

3.3.3 The Creator-Savior's Provisions for Continuity and Order in the Transference of Physical Energy

Although other common denominators exist, the hydrogen atom is the major common denominator for understanding the practical provisions that the Creator-Savior made for transference of energy using ordinary matter in the physically-observable universe, especially as it relates to biological life on Earth.

The majority (approximately 70%) of the Sun in our solar system is composed of hydrogen. This is roughly consistent with hydrogen's percentage of ordinary matter in the entire physically-observable universe (75%). After hydrogen, the remainder (28%) of the Sun is mostly helium. However, even the helium that exists on the Sun is derived from the nuclear fusion of the Sun's hydrogen atoms.

The most abundant form of hydrogen — that is, its most common isotope — consists of one proton and one electron. When hydrogen's proton is denuded of its electron, the resulting charged atom, or ion, is simply a solitary proton. The solitary proton (i.e., an atom of hydrogen without its electron) represents a common denominator for the subatomic structure of all elements in the material universe. The presence or absence of an associated electron (an electron, by the way, has a negligible mass) represents uncharged energy for the hydrogen atom when the electron is present and charged energy for the hydrogen atom when the electron is absent. The uncharged and charged states of hydrogen are represented, respectively, by the common symbols H for one uncharged hydrogen atom and H^+ for one hydrogen ion (i.e., one proton without an associated electron).

The minuscule amount of divine energy in the Creator-Savior's spiritually-observable universe that became altered and bound to form the substance and energy of the physically-observable universe is best (not *solely*) represented by hydrogen in its uncharged state as a neutral hydrogen atom (H) and in its charged state as a hydrogen ion (H^+ or single proton).

In deep space as well as in solar flares, protons (hydrogen atoms denuded of their electrons) are the most abundant type of charged particle. So much radiation is emitted from them that astronauts on the moon in space suits would not survive during solar flares without additional protection.

Photosynthesis is the primary source for the cellular manufacture of food on Earth. As such, photosynthesis is the major generator of usable chemical energy on Earth. During the most common forms of photosynthesis, visible sunlight helps generate hydrogen ions, whose production is ultimately responsible for the synthesis of special hydrocarbons associated with living things (and known as organic molecules). For the most part, the visible sunlight does this by utilizing various specific hydrogen carrier molecules associated with chlorophyll.

All fossil fuels, such as natural gas and petroleum oil (which are derivatives of the organic molecules produced by living things), and all organic molecules in living things are composed of special hydrocarbons. Concerning the latter, hydrocarbons form the basis of carbohydrates, fats, and proteins. When the chemical bonds of any of these special hydrocarbons are broken, they release energy. The more hydrogen atoms per carbon atom in these special hydrocarbons, the more energy is released in uncontrolled oxidation *(for example,* during the spontaneous combustion of hydrocarbons in fossil fuels by fire) as well as in controlled oxidation *(for example,* during the breakdown of hydrocarbons in living cells by enzymes and their associated coenzymes).

Removing the element hydrogen from organic metabolites (intermediates of carbohydrate, fat, and protein metabolism) and passing it along to hydrogen-carrier molecules (and then, eventually, to oxygen) produces chemical energy that is usable by living cells dependent on aerobic respiration. This process is best seen in (i.e., most efficiently shown by) the aerobic cellular respiration that occurs within the mitochondria of most living plant and animal cells. The process is represented by the following general equation:

$$MH_2 + \text{carrier} \rightarrow \text{H-carrier} + H^+ + M + E^0$$

(M represents the metabolite, H represents hydrogen, and E^0 represents emitted energy)

Oxygen is important to the overall process of energy transfer in aerobic respiration, but its primary role is as a final hydrogen acceptor in order to free hydrogen from hydrogen carrier molecules and, thereby, release as much usable chemical energy as possible. Oxidation, or dehydrogenation, of various organic molecules occurs in living cells in the presence of the appropriate dehydrogenases (i.e., specific enzymes that remove hydrogen). In addition to hydrogen, oxygen can be considered a common denominator for the transfer of energy in the physically-observable universe. To be sure, oxygen is the most common — but not the only — hydrogen acceptor in living things.

Although it is an oversimplification, the complex chemical machinery that the Creator-Savior elaborated in both plant and animal cells was made by Him to safely utilize energy associated with hydrogen atoms and their electrons. Thus, the physical food that the Creator-Savior provided to fallen eternal souls in a protoplasmic condition is bound chemical energy that greatly depends on the transfer of energy between hydrogen and hydrogen carriers as well as from energy transported by electrons associated with hydrogen in the safest way possible.

In contrast to energy bound in the form of subatomic particles and atoms in the physically-observable universe, the energy of the spiritually-observable universe is found in the spiritual light, or glory, of the Lord God Almighty, which is: (1) theion-based energy, (2) unbound energy, and (3) energy in its purest state. The spiritual food that the Creator-Savior once provided (and will again provide) to souls with an *astral gelatinous*™ somatic identity is unbound pure energy (i.e., eternal energy, divine energy, or spiritual energy).

Note: If, for some reason, readers or listeners have concluded that the present author is implying that the Creator-Savior is composed of hydrogen atoms and that, when eternal souls are in Heaven, they will be consuming protons and electrons derived from hydrogen atoms, they have missed the mark completely.

Metaphysically speaking, oxygen represents the spiritual energy: (1) that nourished the *astral gelatinous*™ body of original Man, (2) that nourishes the etheric body of saved man, and (3) that will nourish the restored somatic identity of each eternal soul in Heaven.

Metaphysically speaking, oxygen represents the fruit from the Tree of Life in the Garden of Eden. And just as oxygen is produced from water by the leaves of trees on Earth, so, too, is spiritual energy produced from the Spirit — or "Living Water" — of the Creator-Savior. Coincidentally, the "-gen" suffix in the word *oxygen* means *begetter;* thus, in a figurative sense, oxygen as a final hydrogen acceptor is the *begetter* of human life: Just as no human being can live without oxygen, so, too, no spiritual being can live without the Spirit of the Creator-Savior and His divine energy. Of course, the Creator-Savior, and not oxygen, is the real *begetter* of all true life — spiritual life as well as biological life.

Biochemically speaking, oxygen molecules move from one type of protoporphyrin ring to another — *for example,* from the protoporphyrin ring in chlorophyll of cells involved in photosynthesis to the protoporphyrin ring in hemoglobin of red blood cells (i.e., erythrocytes) as well as to the protoporphyrin ring in myoglobin of muscle fibers (i.e., myofibers). Such a movement of oxygen (as well as the movement of oxygen in other similar metalloprotein systems) not only symbolizes the continuity of life in a carbon-based biosphere but also in the metaphysical ecosystem of the spiritually-observable universe, where spiritual energy flows from the Creator-Savior (i.e., represented by the Tree of Life) to His created (i.e., all spiritual beings made in His complete image and perfect likeness). It is this flow that empowers all actions in the spiritually-observable universe.

In summary, the Creator-Savior has provided for a delicate balance in the transfer of physical energy from the Sun to all biological life on Earth. From the present author's perspective, modern biologists who have attempted to simulate the environmental conditions on Earth when biological life originated have underestimated the important role of the Sun and its various types of powerful radiations through the much thinner atmosphere of the Earth at that time. Finally, although all relevant physical factors are important, it was the Creator-Savior only who harnessed what He made available through cosmic evolution to originate and diversify biological life on Earth. Such an origination and diversification took a significant amount of relative time (in the scope of billions of years).

3.3.4 The Importance of Understanding Duration

As indicated previously (in Section 3.3.1), understanding the concept of duration is required to understand the distance between stars or between the Earth and any star other than the Sun relative to: (1) space travel as well as (2) the expansion of the physically-observable universe away from its center. Additionally, understanding the concept of duration is required to understand the establishment and maintenance of ecosystems on Earth over a significant period of time by the Creator-Savior.

Some people might wonder why the Creator-Savior took almost 14 billion years for cosmic evolution, biological evolution, and consciousness evolution when He could have taken one second to hang the stars, create the diversity of biological life that currently exists on Earth, and imbue one select species with the higher order consciousness necessary: (1) to reflect about itself in the physically-observable universe, (2) to articulate those reflections to themselves, and (3) to communicate those reflections amongst themselves from one generation to the next.

Why did the Creator-Savior take so long? First, because the Creator-Savior resides in eternity, He is not constrained by relative space-time except for timelines that He Himself establishes for reasons that only He Himself knows and decides. Second, the Creator-Savior is perfectly knowledgeable about the requirements for delicate balances in the physically-observable universe. Third, the Creator-Savior knows that it takes a long time: (1) to establish local, regional, continental, intercontinental, and marine ecosystems and biomes; (2) to maintain delicate balances within each ecosystem and biome; and (3) to maintain delicate balances between and among ecosystems and biomes within the Earth's entire biosphere.

For example, the Creator-Savior's perfect understanding of delicate balances in ecosystems and biomes is demonstrated by what He shared with the children of Israel concerning the fate of their enemies as they were on the verge of inhabiting the Promised Land:

> {27} I will send My Fear [i.e., My Terror] before you, and I will destroy all of the people to whom you shall come, and I will make all of your enemies turn their backs to you as they flee from you. {28} I will send hornets before you that shall drive out the Hivite, the Canaanite, and the Hittite from before you. {29} But I will not drive them out from before you in one year so that the land not become desolate and the beasts of the field multiply against you. {30} Instead, little-by-little will I drive them out from before you, until you have increased sufficiently in number and are able to inherit the land [that is, possess and inhabit the Promised Land]. [brackets mine]
>
> *Exodus 23:27-30 KJV Paraphrase*

In the establishment and maintenance of ecosystems and biomes on Earth, it is clear that the Creator-Savior understood the necessity for a slow approach in the origin, diversification, and culling of biological life on Earth: Our Creator-Savior wanted to ensure that what He created would be self-sustaining as well as maximally adaptable to myriad physical changes.

To understand more about duration in relationship to cosmic evolution and biological evolution, the present author recommends that students of *intelligent evolution* take the time to study astronomy, geology, and paleontology in courses by the same name or in a planetary science course that includes all three topics. To be sure, a thorough study of geologic time scales will provide students with important insights concerning duration.

3.3.5 The Increase of Biological Diversification and Complexity through the Creator-Savior's Use of Templates

That biological life originated on Earth is, in itself, a strong intellectual proof for *intelligent evolution* because of its improbability. That biological life became increasingly diversified and increasingly complex after it originated on Earth is an even stronger intellectual proof for *intelligent evolution* because such diversification and complexification are also increasingly improbable. However, that the Creator-Savior used anatomic templates to increase biological diversity and complexity on Earth is the strongest intellectual proof of all. (For the sake of clarity, *anatomic templates* here include cellular, tissue, organ, organ system, whole body form, and embryonic templates.) Using anatomic templates to make biological life increasingly diversified and increasingly complex demonstrates that, from the very beginning of creation, the Creator-Savior possessed intention and purpose in cosmic evolution, biological evolution, and consciousness evolution — all of which was, is, and will continue to remain under His control and guidance.

Just as an understanding of the Creator-Savior's transference of the Sun's energy to biological life on Earth provides the student of *intelligent evolution* with templates for understanding general physiology, so does the Creator-Savior's use of anatomic templates permit the student of *intelligent evolution* to understand: (1) the existence of variations in any one group of organisms, (2) diversification of appearance in the shared basic anatomy of that group, (3) various adaptations that have occurred within the entire group over time, and (4) whether or not specific members of that group have continued to survive and thrive or become extinct.

Increasing the diversity and complexity of biological life on Earth not only refers to the interrelationships and interdependencies of groups of organisms in Earth's various ecosystems and biomes but also: (1) refers to some organisms from the fungal, plant, and animal kingdoms in individual ecosystems closely resembling one another; (2) refers to the general recapitulation of phylogeny by ontogeny; and (3) refers to progressive cerebralization in the seven *ascending* classes of the vertebrate subphylum[28] as well as in the respective *ascending* subclasses, orders, families, genera, and species of the mammalian class to finally produce *Homo sapiens* — the only species capable of: (a) thinking for itself; (b) reflecting on who it is; (c) knowing that it knows; (d) developing spoken and written language; (e) transmitting information from one generation to the next; (f) worshiping the Creator-Savior; and (g) housing fallen eternal souls to provide them with opportunities for their salvation by accepting Christ Jesus for their restoration to *Paradise.*

For the sake of clarification concerning taxonomic classification in the previous paragraph, "ascending" refers to groups of organisms that are "increasingly complex," "higher," "sequential" (in terms of the initial emergence of a group on Earth relative to the emergence of other related groups), and, ironically, "descending" (based on the progression of descendants or the succession of organismic groups). When groups of organisms in relationship to one another are referred to as "inferior" and "superior," "lower" and "higher," or "earlier" and "later," the reference is to where they appear in their phylogenetic arrangement on a biological tree of life and not to their relative worth, value, or importance — although, in one way, earlier groups are more important than later groups because the later groups would not exist without the earlier ones existing first. If anyone ever prays to the Lord God Almighty to get rid of gnats and flies because one thinks that they are a nuisance and, therefore, have no worth, value, or importance, I believe that our Creator-Savior would reply (as He denies that

[28] The seven ascending classes of the vertebrate subphylum include: (1) Class Agnatha (jawless fishes), (2) Class Chondrichthyes (cartilaginous fishes), (3) Class Osteichthyes (bony fishes), (4) Class Amphibia (amphibians), (5) Class Reptilia (reptiles), (6) Class Aves (birds), and (7) Class Mammalia (mammals).

request): "You certainly do not understand what you are asking. So that you never make the same mistake again, you need to study about how I create and maintain delicate balances in ecosystems and biomes. Unfortunately, or fortunately, for you, flies and gnats are necessary for your overall survival!"

Just as the Creator-Savior uses essentially the same molecular, or physiologic, templates for energy transfer throughout the Earth's biosphere (in the similar cellular enzyme kinetics and associated electron transport systems in the cells of many individual living organisms), the Creator-Savior also uses essentially the same anatomic templates to morph the morphology of new species from previously existing species that have either become extinct or continued to exist. Of course, the Creator-Savior created variations and adaptations within all groups of organisms over a considerably long period of time to enable the individual ecosystems and biomes of the Earth's biosphere to get to where they are now — stable yet responsive to environmental changes and shifts. (To be sure, human beings continue to negatively impact biosphere stability.)

As an example of the Creator-Savior's use of anatomic templates, the anatomy of a hummingbird is essentially the same as the anatomy of a penguin as well as the anatomy of an ostrich because their individual anatomic forms are based on the same general anatomic templates (including cell, tissue, organ, organ system, whole body form, and embryonic templates) from a common ancestor. Any differences between and among hummingbirds, penguins, and ostriches are due to: (1) variances in their avian genetic codes; (2) variances in the concentrations, kinds, and proportionalities of their avian hormones; and (3) variances in their individual avian metabolisms — all in relationship to their individual group's survivability, thrivability, and sustainability in their respective habitats.

Through the direction and guidance of the Creator-Savior, organisms have the innate tendency to change by responding to changing environmental factors as well as by re-organizing their own DNA and RNA. Indeed, internal mechanisms existed during the six "days" of creation that drove *intelligent evolution* not only to produce *Homo sapiens* but also to support *Homo sapiens* in very special ways. Diversification of biological life exists: (1) to support *Homo sapiens* practically with food, fuel, and durable protection; (2) to enable *Homo sapiens* to appreciate and marvel in the existence of diverse biological forms and functions and their interrelationships; (3) to provide *Homo sapiens* with an endless source of study for the discovery of important principles and applications; and (4) to evoke in *Homo sapiens* a recognition of metaphysical truths. To be sure, *intelligent evolution* proves that the Creator-Savior has an expansive imagination as well as an insatiable intellectual curiosity. The Creator-Savior even imparted that imagination and intellectual curiosity to us in our immortal state because He originally made us in His complete image and perfect likeness. And, although we have fallen from immortality to mortality, we still retain many vestiges of the attributes with which we were originally endowed.

3.3.5.1 Mimicry and Camouflage

We are asked by evolutionists to believe in the existence of "selective pressures" on specific species either (1) for one species to closely resemble another species through its own evolution or (2) for two species to closely resemble each other through their coevolution. In other words, we are asked to believe that the existence of two species closely resembling each other is due to accidentally occurring variations that either (1) favor the survival of one species over the other or (2) have mutually beneficial impacts on the survival of both species.

For the sake of clarity, the use of the word *evolutionists* in the preceding paragraph specifically refers to people who do not incorporate a concept of the Creator-Savior into their personal theory or paradigm of evolution: (1) because they do not believe that a Creator-Savior is involved in evolution, (2) because they do not believe that a Creator-Savior exists, and/or (3) because they believe that the Creator-Savior has no place in a discussion about evolution, regardless of whether He exists or not.

In some cases, the enhanced survival of one species based on mimicry is because the species is less visible to, or identifiable by, its natural predators. Indeed, it is true that some species have a survival advantage due to mimicry, but it is not true that the appearances of the species are the result of chance as most people think of chance. Instead, in keeping with pure reason and metaphysical understanding, one species closely resembling another species is the effect of *directed chance* and *intelligent evolution* by *the Supraconsciousness of the Creator-Savior*.

Mimicry and camouflage are important clues to members of *Homo sapiens* that there is a Creator-Savior. Mimicry and camouflage also provide important clues of the Creator-Savior's expansive imagination, intellectual curiosity, and artistry in His use of biological templates across specific kingdoms as well as across specific groups within one kingdom.

Because this book has no photographs, the present author directs its readers to look up relevant examples of two different species resembling one another on the internet and/or at a library with paper-printed books. Following are ten examples of mimicry or camouflage that are beyond explanations from calculable chance mutations of DNA or RNA:

1. bee orchids: Not only do the flowers of this plant resemble the insects that pollinate them, they also produce scents that mimic the sexual pheromones of the specific pollinating insects in order to attract them.

2. angler fish: An angler fish displays an appendage that resembles a small fish in order to attract other predators for the angler fish to ingest. This is an example of aggressive mimicry where the predator develops a significant advantage over its prey.

3. leaf insects: These represent a range of insects that resemble the leaves of various plants. The "leaves" displayed by the insects include specialized venation, serration, perforation, movement, and coloration. (Concerning movement, the swaying movements of some leaf insects resemble leaf movements from breezes and winds. And, concerning coloration, some leaf insects are the same color of leaves because of the actual leaf pigments they have ingested.)

4. walkingsticks: There are more than 3,000 species of insects that resemble twigs or branches up to two feet in length (a twig is a small tree branch). Their multiple "sticks" are complete with structures that resemble the spines, nodes, and internodes of branches. Moreover, some species of walkingsticks also have uncommon regenerative abilities when an appendage or antenna is broken at a "node."

5. hawkmoth caterpillars: Some of these caterpillars pull in their appendages and head at the same time that they puff out their front end to resemble a snake's head in order to intimidate looming predators.

6. mimic octopuses (mimic octopi/octopodes): A mimic octopus can change its coloration and movement to mimic up to fifteen different species in order to more successfully blend in with its immediate environment to avoid predation as well as capture prey. An individual mimic octopus can resemble various fish, crabs, algae-encrusted rocks, corals, jellyfish, sea snakes, sponges, tube-worms, and tunicates.

7. orchid mantises: An orchid mantis resembles an orchid flower in order to lure insects, such as bees and butterflies, which it then captures and digests.

8. leaf-tailed geckos: The head, thoracic, abdominal, and tail regions of these lizards closely resemble true leaves in size, shape, venation, and coloration, including variegation.

Depending on species and sex, some of these geckos resemble fresh leaves and others resemble older fallen leaves that have been partially fed upon by insects.

9. caterpillars: Some moth and butterfly caterpillars (like the Celerio Sister, Adelpha serpa selerio, and saterniid caterpillars) have: (a) specialized protrusions that resemble plant prickles, spines, barbs, or needles; and (b) unusual color patterns and textures that resemble forest ferns, fungi, mosses, slime molds, and/or lichens.

10. katydids: A wide range of mimicry is seen in various species of these grasshoppers.
    Readers of *Intelligent Evolution* are encouraged to use search engines on the internet to find digital images of additional examples that represent mimicry and camouflage.

3.3.5.2 Recapitulation and Biogenetic Law

The simplest way to summarize recapitulation and biogenetic law so as not to alarm both biologists and creationists is: (1) the earliest stages of embryonic development in vertebrates resemble a number of stages in invertebrate development; (2) the embryos of most vertebrates greatly resemble the embryos of many other vertebrates; (3) the developing embryos of most vertebrates possess significant morphological similarities; (4) the embryos within one vertebrate class closely resemble each other; (5) the embryos from one vertebrate class somewhat resemble the embryos from the six other vertebrate classes (see Footnote 28 for the seven vertebrate classes); and (6) in an ascending order from a phylogenetic standpoint, embryos from the seven vertebrate classes demonstrate shared similarities based on general morphological templates.

Because the topic here was covered to the present author's satisfaction elsewhere in this book, the reader is asked to reread the subsection entitled *Haeckel* in Section 2.5.4.4.

To be sure, comparative anatomy of vertebrates at their embryonic levels provides many insights of the Creator-Savior's use of morphological templates as well as His re-use of them with modifications to provide us with a primer picture book understanding (i.e., a kinetoscopic understanding) of succession in phylogenetic developments. Although biologists and creationists do not believe that the theory of recapitulation provides sufficient evidence to support biological evolution, many aspects of the theory of recapitulation work well for a metaphysical understanding of *intelligent evolution.*

Metaphysically speaking, the embryos of organisms existing today are living fossils of past biological life. Unfortunately, because animal embryos that have no bone, calcified cartilage, enamel, dentin, or keratin do not fossilize, they are not found as fossils in either amber or ice (polar, glacier, marine, alpine, or hailstone ice). Thus, we may only study animal embryos of the past by studying animal embryos of the present.

3.3.5.3 Consciousness and Cerebralization

The biological ascent of consciousness is reflected in reflection, especially self-reflection — signaled by such questions as: "Who am I?" "Why am I here?" "What is my purpose for *being?"* Increased cerebralization is very much a part of the biological ascent of consciousness. Without the embryologic development of the forebrain and its associated cerebral convolutions, *Homo sapiens* would show very few functional differences from the chimpanzee, *Pan troglodytes,* with whom it shares approximately 98.6% of its DNA. Although chimpanzees demonstrate communication, tool use, culture, and individual personalities, the chimpanzee species is still inadequate for the habitation of fallen eternal souls, who require more advanced cerebralization to perform activities associated with the capabilities listed as *a* through *f* in the third introductory paragraph of Section 3.3.5. To understand more about cerebralization and cranial capacities of evolving prehistoric mankind, the present author recommends that students of *intelligent evolution* take the time to study anthropology as well as certain aspects of paleontology that relate especially to humankind.

Does consciousness evolve or does life on Earth evolve to express more already-existing consciousness than it did previously? In one way, it is not really consciousness that evolves but the cerebral means to understand and express consciousness — as well as understand what *self* is — that evolves. However, in another way, consciousness can evolve not only in one person but also in a group of people and even in the entire human race. Degree of awareness of one's *self* in relation to other human beings as well as to invisible entities that also possess consciousness is an important feature of evolving consciousness. So, distinctions can be made between human consciousness and spiritual consciousness in a human being. Evolving human consciousness primarily has applications to being human. In contrast, evolving spiritual consciousness primarily has applications to eternal *being*. However, these two types of consciousness can intersect when one learns a human lesson that has spiritual, or metaphysical, applications.

People who reflect about themselves without becoming self-absorbed, self-obsessed, egotistical, arrogant, haughty, and paranoid can become better reflectors of the Creator-Savior's complete image and perfect likeness. Unfortunately, narcissistic personality disorder (NPD) can be traced back to Lucifer, the first spiritual being who fell from immortality to mortality. (Please remember that, although corporeality is a part of mortality, corporeality does not constitute the entirety of mortality. Mortality is, first and foremost, a fallen spiritual state of being.)

Because of their inherited iniquity, human beings are prone to narcissistic personality disorder. Our predisposition to NPD is why the Creator-Savior permits (1) "thorns in the flesh" from messengers, or angels, of Satan and (2) unclean spirits who attack us: Their presence helps to keep us living in a state of perpetual contrition because their presence forces us to make, minute-by-minute, decisions either to indulge the temptations with which they bait and bombard us or to reject what they offer in order to please our Creator-Savior. In other words, in the human condition, we are constantly being given opportunities to choose good over evil.

Many people are aware of actual demons in their lives. The present author is one of those people. If you, too, are such a person, please know that eventually the Creator-Savior will remove such a *thorn in the flesh* from you either (1) because you have continued to make the right decisions for a period of time that He has predetermined or (2) because you have become so physically ill that what the demons have to offer has absolutely no appeal to you. (Yes, a debilitating illness can be a blessing to some human beings for it is true, indeed, that "the person who suffers in the flesh ceases from sin." *1 Peter 4:1 KJV Paraphrase)*

>>>>><<<<<

In closing, study in many areas will provide students of *intelligent evolution* with important insights about comparative anatomy that thoroughly demonstrate the existence of evolving morphological templates. Templates from the seven vertebrate classes especially exemplify the coherence of all vertebrate classes relative to *intelligent evolution. For example,* of the approximately 50,000 species that possess a vertebral column, the overwhelming majority possess seven cervical vertebrae. Metaphysically speaking, the various vertebrate templates overlie electromagnetic templates that exist in the spiritually-observable universe, specifically in created spiritual beings who reside in the Supraconsciousness of the Creator-Savior (i.e., *the Mind of God).*

Afterword to Volume Two

The author of this book believes that truly educated people should be able to cogently argue on behalf of any position on either side of the aisle while keeping their own biases and conclusions to themselves. However, he also believes that there is a time and a place to share one's personal views with conviction. For the present author, now is the time and this book is the place to share his views concerning *creation* and *salvation* as they relate to *intelligent evolution*.

The original creation of all spiritual life was an act by the Creator-Savior in an instant in eternity. The creation of all biological life was an act by the Creator-Savior over billions of years, and *intelligent evolution* is the process by which that act was achieved. Finally, the re-creation of spiritual life for fallen eternal souls was also an act by the Creator-Savior, and the shedding of blood by Christ Jesus on the cross of his crucifixion is the process by which that act was achieved. To be sure, the cross of Christ's crucifixion is where eternity intersects temporality and where the Creator-Savior has formed a temporary rift in mortality for fallen souls to return to *Paradise* as the immortal beings they once were.

For the author, the creation of biological life through *intelligent evolution* and the re-creation of spiritual life through *salvation* are the two most important unifying laws that attest to the simplicity, complexity, and power of Christ Jesus as well as to the essence of life — biological life as well as spiritual life. It does not disappoint the present author to know that few people during his lifetime will share the glimpse of truth that he has had in understanding *intelligent evolution*. Why is he not disappointed? If people have not recognized that the Creator-Savior was in human flesh as the *only-begotten* Son of God, and if people still cannot accept the gospel message of Christ Jesus, then how can the author be disappointed that people will not recognize the truth behind his paradigm of *intelligent evolution?*

The author has begun his seventy-fourth year on Earth with the 2021 edition of *Intelligent Evolution.* Although he has published this book, authors are never really finished with their work. Indeed, authors often return to their literary efforts to touch up, refine, and polish what they have written. For the remainder of his life, the present author intends to revisit all of his books for continued editing, revising, and reformatting. However, the day will come when he no longer can do that. For that reason, the author is pleased to report that, as of this very day, he believes that he has done his best in the writing of his books with the nurturance, training, education, mentoring, and discipline that he was blessed to receive throughout his life. The author is grateful to the Creator-Savior for His ongoing guidance and leadership as well as for the initiative, motivation, ideas, insights, and energy given to him daily to complete the work with which he was tasked. The author is grateful for this spoken assurance from the Creator-Savior concerning the future availability of his books: "I will make sure they are published." And the author is grateful for this spoken review that he received from Heaven upon completing *Intelligent Evolution:* "This book presents evidence of God's intellect and that of the sower's." (In the final analysis, approval and validation of one's work comes first and foremost from Heaven.) It is the Creator-Savior alone who ordains fulfillment with real purpose.

Because it is the Creator-Savior who opens doors that no one can close, and because it is the Creator-Savior who closes doors that no one can open, the author entrusts all of his books to the Creator-Savior to do with as He will. It is with great faith, hope, and love that the author finishes *Intelligent Evolution* as well as closes this specific chapter of his life.

Bibliography: Volumes One and Two

Aristotle. *Metaphysics: Translated with an Introduction by Hugh Lawson-Tancred.* New York: Penguin Books, 2004.

Aristotle. *Physics: A New Translation by Robin Waterfield.* Oxford: University Press, 2008.

Berry, George Ricker. *The Interlinear Literal Translation of the Hebrew Old Testament.* (Reprinted from the 1897 Edition). Grand Rapids: Kregel Publications, 1979.

BSCS Biology: A Molecular Approach (Blue Version 9th Edition, Annotated Teacher Edition). Ohio: Glencoe/McGraw Hill, 2006.

Bullinger, E. W. *Figures of Speech Used in the Bible.* (Reprinted from the 1898 Edition). Grand Rapids: Baker Book House, 1968.

Bullinger, E. W. *The Companion Bible (Facsimile Edition).* Grand Rapids: Kregel Publications, 1922.

Darwin, Charles. *The Origin of Species by Means of Natural Selection of the Preservation of Favoured Races in the Struggle for Life.* New York: Signet Classics by New American Library, a division of Penguin Group (USA) Inc., 2003.

Dickey, Adam H. *Memoirs of Mary Baker Eddy.* Santa Clarita: The Bookmark, 2002.

Ebert, James D. and Ian M. Sussex. *Interacting Systems in Development (second edition).* New York: Holt, Rinehart and Winston, Inc., 1970.

Eddy, Mary Baker. *Complete Concordance to Miscellaneous Writings and Works other than Science and Health.* Boston: Trustees under the Will of Mary Baker G. Eddy, 1915.

Eddy, Mary Baker. *Concordance to Science and Health with Key to the Scriptures.* Boston: Trustees under the Will of Mary Baker G. Eddy, 1933.

Eddy, Mary Baker. *Prose Works other than Science and Health with Key to the Scriptures.* Boston: The First Church of Christ, Scientist, 1953.

Eddy, Mary Baker. *Science and Health with Key to the Scriptures.* Boston: Christian Science Board of Directors, 1906.

Gishlick, Alan D. "Haeckel's Embryos" in *Icon of Evolution? Why much of what Jonathan Wells writes about evolution is wrong*, pages 26-40. Oakland: National Center for Science Education, 2003.

Haeckel, Ernst. *Anthropogenie oder Entwicklungsgeschichte des Menschen.* Leipzig: Verlag von Wilhelm Engelmann, 1874.

Haeckel, Ernst. *The Evolution of Man: A Popular Exposition.* Volumes One & Two. New York: D. Appleton and Company, 1897.

Haushalter, Walter M. *Mrs. Eddy Purloins from Hegel.* Boston: A. A. Beauchamp, 1936.

Hawking, Stephen. *A Brief History of Time.* New York: Bantom Books, 2017.

Hawking, Stephen. *Brief Answers to the Big Questions.* New York: Bantom Books, 2018.

Hegel, Georg Wilhelm Friedrich. *Die Phänomenologie des Geistes (1807).* Project Gutenberg eBook, 19 June 2012 <www.gutenberg.org/catalog/world/readfile?fk_files=1464225>.

Kant, Immanuel. *Prolegomena to any Future Metaphysics Which Will Be Able to Come Forth as Science (translation from 1783 edition)*. New York: The Liberal Arts Press, 1950.

Kant, Immanuel. *Prolegomena zu einer jeden künftigen Metaphysik die als Wissenschaft wird auftreten können (Erstdruck: Riga 1783)*. Berliner Ausgabe, 2 Auflage, 2013.

Kant, Immanuel. *The Metaphysical Foundations of Natural Science*. Lexington: Translated by Ernest Belfort Bax, 2015.

Lamarck, Jean Baptiste. *Zoological Philosophy: An Exposition with regard to the Natural History of Animals*. London: MacMillan and Co., Limited, 1914.

Levinson, Gene. *Rethinking Evolution: The Revolution That's Hiding in Plain Sight*. World Scientific: New Jersey, 2019.

Mary Baker Eddy Mentioned Them. Boston: The Christian Science Publishing Society, 1961.

Nelson, David L. and Michael M. Cox. *Lehninger Principles of Biochemistry*. New York: W. H. Freeman, Fifth Edition, 2008.

Pearson, Joseph Adam. *As I See It: The Nature of Reality by God*. Michigan City: Christ Evangelical Bible Institute, 2015.

Quimby, Phineas Parkhurst. *The Quimby Manuscripts (1846-1865)*. London: Forgotten Books (Classic Reprint Series), 2015.

Richards, Robert J. "Haeckel's embryos: fraud not proven" in *Biology and Philosophy* (2009) 24:147–154.

Singer, Isadore. *The Jewish Encyclopedia: 1906 Edition*. New York: Funk and Wagnalls, 1906.

Strong, James. *Strong's Exhaustive Concordance of the Bible.* Nashville: Crusade Bible Publishers, Inc., 1890.

Teilhard de Chardin, Pierre. *Christianity and Evolution: Reflections on Science and Religion.* Orlando: A Harvest Book · Harcourt, Inc., 1974.

Teilhard de Chardin, Pierre. *Le phénomène humain.* Paris: Editions du Seuil, 1955.

Teilhard de Chardin, Pierre. *The Phenomenon of Man.* New York: Harper Perennial Modern Thought, 2008.

Thorne, Kip. *The Science of Interstellar.* New York: W. W. Norton & Company, 2014.

Walvoord, John F. and Roy B. Zuck (editors). *The Bible Knowledge Commentary: An Exposition of the Scriptures by Dallas Seminary Faculty [New Testament Edition].* Elgin: David C. Cook, 1983.

Walvoord, John F. and Roy B. Zuck (editors). *The Bible Knowledge Commentary: An Exposition of the Scriptures by Dallas Seminary Faculty [Old Testament Edition].* Colorado Springs: Cook Communications Ministries, 2000.

Vine, William E., Merrill F. Unger, and William White. *Vine's Complete Expository Dictionary of Old and New Testament Words.* Nashville: Thomas Nelson, Inc., 1985.

Webster, Noah. *Noah Webster's First Edition of An American Dictionary of the English Language (Facsimile Edition).* Anaheim: Foundation for American Christian Education, 1967.

Zondervan Parallel New Testament in Greek and English. Grand Rapids: Zondervan Bible Publishers, 1975.

Books by the Author

As I See It: The Nature of Reality by God by Rev. Joseph Adam Pearson, Ph.D., Christ Evangelical Bible Institute, Copyright 2022. ISBN 978-0615590615.

Classroom Version of As I See It: The Nature of Reality by God by Rev. Joseph Adam Pearson, Ph.D., Christ Evangelical Bible Institute, Copyright 2022. ISBN-13: 978-1734294705.

God, Our Universal Self: A Primer for Future Christian Metaphysics by Rev. Joseph Adam Pearson, Ph.D., Christ Evangelical Bible Institute, Copyright 2020. ISBN 978-0985772857.

Divine Metaphysics of Human Anatomy by Rev. Joseph Adam Pearson, Ph.D., Christ Evangelical Bible Institute, Copyright 2021. ISBN 978-0985772819.

Hello from 3050 AD! by Rev. Joseph Adam Pearson, Ph.D., Christ Evangelical Bible Institute, Copyright 2021. ISBN 978-0996222402.

Christianity and Homosexuality Reconciled: New Thinking for a New Millennium! by Rev. Joseph Adam Pearson, Ph.D., Christ Evangelical Bible Institute, Copyright 2021. ISBN 978-0985772888.

The Koran (al-Qur'an): Testimony of Antichrist by Rev. Joseph Adam Pearson, Ph.D., Christ Evangelical Bible Institute, Copyright 2020. ISBN 978-0985772833.

Telugu Version of Quran: Testimony of Antichrist by Rev. Joseph Adam Pearson, Ph.D., Christ Evangelical Bible Institute, Copyright 2020. ISBN 978-0996222457.

Urdu Version of Quran: Testimony of Antichrist by Rev. Joseph Adam Pearson, Ph.D., Christ Evangelical Bible Institute, Copyright 2021. ISBN 978-0996222440.

Revelation of Antichrist by Rev. Joseph Adam Pearson, Ph.D., Christ Evangelical Bible Institute, Copyright 2021. ISBN 9780996222488.

Intelligent Evolution by Rev. Joseph Adam Pearson, Ph.D., Christ Evangelical Bible Institute, Copyright 2022. ISBN 978-0996222426.

The Biology of Psychism from a Christian Perspective by Rev. Joseph Adam Pearson, Ph.D., Christ Evangelical Bible Institute, Copyright 2020. ISBN 978-0996222464.

The Threeness of God by Rev. Joseph Adam Pearson, Ph.D., Christ Evangelical Bible Institute, Copyright 2021. ISBN 978-1734294729.

The author may be contacted
at
drjpearson@aol.com
or
drjosephadampearson@gmail.com

Visit the author's legacy websites
at
www.dr-joseph-adam-pearson.com
or
www.christevangelicalbibleinstitute.com

My Eternal Now

When my soul unfurls its wings and wafts heavenward,
The memories that it brings will be pleasant ones,
Those of this life's stings with my human cocoon lie,
What I once wore as drab attire sloughed off,
My shimmering self now sings only of God's own things.
Indeed, life is precious in all its forms:
From the supraplasmic shapes of His spiritual creation
To the protoplasmic ones of His earthly design
I like them all but I like Heaven's best.
There, there is no time but the present rest.

Rev. Joseph Adam Pearson, Ph.D.
May 4, 2022

Made in the USA
Monee, IL
10 November 2026

23dc6312-07e0-4484-bd7b-9d8a2b36cd19R01